Ornamental Gardening and Landscaping

nipa
Browse Subject
Agriculture Sciences (679)
Animal Husbandry and Veterinary Sciences (761)
Climate Change and Environmental Sciences (63)
Community Science (Home Science) (39)
Competitive Examinations and MCQ's (57)
Dairy Technology (91)
Fisheries Sciences (26)
Food Science and Technology (155)
Geological Sciences (50)
Horticultural Sciences (342)
Plant Sciences (191)
Skill and Entrepreneurship Development (47)
eBooks / eChapters / Articles
eBooks
Explore Now
An e book is an electronic book available in various digital formats, can be accessed from anywhere, anytime on desktops/laptops/smart phones and tablets connected to internet.
scan for catalogue
Browse, Search, Read & Buy...
Subject Catalogues
eChapters
Publishers
Print Books
Forthcoming
eBooks
New Titles
NIPA
GENX
ONLINE RESOURCES
Publishing Books and Journals
Ebooks and e articles, Current Affairs
Reasoning, Logic and Aptitude
Competitive Examination Preparation
Language Learning and Development Programme
Document Quality Checker and Improvement Tool
Effective Public Speaking, Presentation and Interpersonal Skills
Online Programmes for Professional Development
Personality Development and Human Values
PAY USING
UPI
PayPal

Ornamental Gardening and Landscaping: A Guide

Authors

M. Kannan

P. Ranchana

S. Vinodh

Department of Floriculture and Landscaping
Tamil Nadu Agricultural University
Coimbatore - 641 003, Tamil Nadu, India

NEW INDIA PUBLISHING AGENCY

New Delhi – 110 034

NEW INDIA PUBLISHING AGENCY
101, Vikas Surya Plaza, CU Block, LSC Market
Pitam Pura, New Delhi 110 034, India
Phone: +91 (11)27 34 17 17 Fax: +91(11) 27 34 16 16
Email: info@nipabooks.com
Web: www.nipabooks.com

Feedback at feedbacks@nipabooks.com

ISBN : 978-93-85516-68-9

Composed and Designed by NIPA

Dr. B. Hemla Naik
Ph.D. (Hort.); IPGDH (Israel); elearning (USA)
Professor and Head (Hort.) cum Nodal officer (PME)
University of Agricultural and Horticultural Sciences
Shivamogga - 577 225, Karnataka (India)

Foreword

Gardening which was only an art and science in the earlier days has now emerged as a huge industry. With the importance and need of gardening in improving and conserving the environment being strongly felt now, the concept of landscaping and gardening is growing rapidly. Ornamental gardening and landscaping has expanded as a multi-faceted industry encompassing activities such as propagation and rearing of ornamental plants, landscaping, production of growing media, production of plants in pots and other accessories, etc., generating huge employment opportunities and simultaneously promoting activities that would improve the environment.

By virtue of its importance in agribusiness, research and education, ornamental gardening and landscaping has reached top status. ICAR, New Delhi is conducting examinations like ARS, NET, Senior Research Fellowship, and Junior Research Fellowship separately on Floriculture. This book includes objective type of questions on all aspects of ornamental gardening & Landscaping and act as a ready reckoner for the students to update their knowledge in this subject. This book will be more useful for the students who are appearing the ICAR examinations (JRF, SRF, NET, ARS) and competitive examinations for State Horticulture/ Agriculture Department officers being conducted by state governments and Agricultural officers in Banks.

Though there are many books on 'Ornamental Gardening' which are being published since longtime, a comprehensive and compressed book to be readily used for competitive examinations by students is scanty. Hence, I am sure that this book on 'Ornamental Gardening and Landscaping : A Guide' would be definitely useful to them to overcome all competitive examinations related to Ornamental gardening and Landscaping. I wholeheartedly appreciate and congratulate the efforts taken by the authors *viz.*, Dr. M. Kannan, Dr. P. Ranchana and Dr. S. Vinodh to bring out this book. Further, this book will also be useful for the teachers, researchers, Landscape consultants, architects and garden lovers also involved in 'Ornamental Gardening and Landscaping' to refresh their knowledge and get updated.

(B. HEMLA NAIK)

Preface

Gardens have become essential to modern civilization. Home gardening is an integral part of family life and it lends itself for the family to relax and enjoy the nature and beauty of garden. It also helps to develop some useful hobby and the 'skill' in fresh and dry flower arrangements. In the recent years, landscape architecture has come to play a vital role in the development of home, public institutions, public places like bus stand, railway stations, highways in towns and cities.

Landscaping the roadsides in cities and towns is gaining importance in tropical countries like India as they provide shade and beauty to that area. In foreign countries landscaping of homes and public buildings is an important art and there are professional and qualified landscape architects who do the job. This field of gardening and landscaping is slowly gaining momentum in our country especially in big cities. The Government has also established many parks and a few Botanic gardens in the country.

Competitive examinations nowadays have become a procedure for the recruitment, admission for higher studies in the respective field. Questions asked in these examinations are of objective type, many in numbers and varied in nature, which require quick answering within a limited period of time. ICAR, New Delhi has been conducting several competitive examinations in the field of JRF, SRF, NET, ARS etc. Moreover, several state agricultural and horticultural universities in India are also conducting entrance examinations for admission leading to higher studies.

We have made an attempt to compile all the basic and latest information in "Ornamental Gardening and Landscaping : A Guide" in the form of a question bank. This will be highly useful for the students who are preparing for JRF, SRF, NET, ARS of ICAR and other examinations leading to admission for higher studies and for recruitment. At this juncture, we would like to thank all the staff and students of Department of Floriculture and Landscaping, Horticultural College and Research Institute, TNAU, Coimbatore who extended cooperation during preparation of this book. We express our deep sense of gratitude to New India Publishing Agency, New Delhi for taking up this assignment within a short span of time.

Authors

Contents

Chapter **1**

Principles of Gardening

Principles to create effects

Balance

It is a visual equilibrium of different garden elements.

Formal Garden : Balance is achieved by positioning plants, other landscape objects at equal distance from a real (or) imaginary plane (or) axis.

Informal Garden : Balancing is mainly a neutralizing effect. (eg.) Opposite colours in the colour wheel, contrasting texture, different numbers, shapes, forms, line and pattern are cleverly employed to neutralize one another.

Proportion

It refers to the share of the different parts (or) components to the whole. It refers to all ingredients being mixed in right measure, not essentially in equal measure.

Harmony

It is the pleasing effect due to an apt arrangement & collation of the various garden features. Every part of the landscape must unobtrusively merge into a whole.

Rhythm: It is measured as a cyclic repetition which denotes occurance of things one after another.

Repetition

This is having direct relation with unity. Repetition of uniform elements having similar shape and colour creates unique harmonious effect.

Unity

Aesthetic principles of balance, rhythm, proportion and harmony have to be merged to give unity to the composition. Unity should have a central theme for the composition.

Perspective

The visual phenomena of shrinking size and converging lines in a three-dimensional space are known as perspective

Vista

Vista is the view of an object, focal point through a long narrow passage. They generate interest and curiosity. Statues, fountains, a specimen plant, etc. can be employed in a garden to create a vista.

Prospect

It can be created in landscape garden by adopting proper scale and proportion. It is the view of sceneries, natural (or) man-made, through an opening such as window (or) a gap in the foliage of trees.

Accent and Specimen trees

Accent, specimen trees and contrast show dominance over others. These are uncompromising aesthetic appeal towards themselves, among the many lesser plants in groups.

Restraint

In designed landscape, avoid over use of anything including plant grouping. Keep all the features within bounds (or) use restraint to achieve the best results.

Scale

It is a relative dimension. Appropriate adoption of scales and proportionate measurements are the success of imitative naturalistic garden art.

Simplicity

This criterion emphasizes preparing simple design as per style & theme. Crowding of objects, plants, decors should be avoided in order to give prominence to each feature.

Unity

Aesthetic principles of balance, rhythm, proportion and harmony have to be merged to give unity to the composition. Unity should have a central theme for the composition.

Perspective

The visual phenomena of shrinking size and converging lines in a three-dimensional space are known as perspective

Vista

Vista is the view of an object, focal point through a long narrow passage. They generate interest and curiosity. Statues, fountains, a specimen plant, etc. can be employed in a garden to create a vista.

Prospect

It can be created in landscape garden by adopting proper scale and proportion. It is the view of sceneries, natural (or) man-made, through an opening such as window (or) a gap in the foliage of trees.

Accent and Specimen trees

Accent, specimen trees and contrast show dominance over others. These are uncompromising aesthetic appeal towards themselves, among the many lesser plants in groups.

Restraint

In designed landscape, avoid over use of anything including plant grouping. Keep all the features within bounds (or) use restraint to achieve the best results.

Scale

It is a relative dimension. Appropriate adoption of scales and proportionate measurements are the success of imitative naturalistic garden art.

Simplicity

This criterion emphasizes preparing simple design as per style & theme. Crowding of objects, plants, decors should be avoided in order to give prominence to each feature.

Natural transition

This refers to gradual changes in terms of plant height, colour, forms & size besides arrangement in ascending (or) descending patteren.

Colour

Colour of the flowers and leaves constitute a critical factor for whole design of the gardens as it creates attraction. Primary colours (violet, blue, green, yellow, orange and red) as well as secondary colours are to be used either in contrast or harmonious way depending upon the availability. Generally bright colours are put close to the visitors while cool colours at distant part of the garden.

Texture

The surface character of the garden unit is generally referred to as texture. It is expressed from fine to medium to coarse. It is used to emphasizes or minimizes architextural lines.

Form

It is the result of the total mass of any object or a plant in landscape design. It is important in creating a particular design in a landscape.

Chapter 2

Famous Gardens of India

I. General information

1. Lalbagh garden is located at Bangalore
2. Founder of Lalbagh garden is Hyder Ali and his son Tipu Sultan
3. Total area of Lalbagh garden is 50 hectares and developed during 1760
4. Brindavan garden is located at Mysore
5. Government botanic garden, Ooty was established during 1848
6. Total area of Ooty botanic garden is 20 hectares
7. The sections of Ooty botanic garden are six
8. Six sections of Ooty botanic garden are
   - Lower garden
   - Bandstand
   - Conservatory, bog garden and tennis court
   - Terrace with fountain and picnic spots
   - New bandstand and its surroundings
   - Nursery area
9. An important essential oil yielding plant introduced by Ooty botanical garden is scented geranium
10. Sims park is located at Coonoor
11. Sims park was established during 1874
12. Founder of Sims park is J.D. Sim
13. Total area of Sims park is 15 hectares

14. Botanic garden, Coimbatore was established during 1908
15. Coimbatore botanic garden was established by Department of Agriculture, Madras
16. The Bryant park is located at Kodaikanal
17. Total area of Bryant park is 10 hectares
18. The Indian botanic garden is located at Shibpur, Calcutta
19. The Indian botanic garden is located opposite side of the river Hoogly
20. The Indian botanic garden was established during 1787
21. Total area of Indian botanic garden is 150 hectares
22. Most attractive features of Indian botanic garden are 200 year old banian tree, palmatum with royal palm, avenue near the river gate and giant lily
23. A botanic garden of India which ranks among the great botanical gardens of the world is Indian botanic garden, Shibpur, Calcutta
24. Llyod botanic garden is located at Darjeeling, West Bengal
25. Llyod botanic garden was established during 1878
26. Total area of Llyod botanic garden is 24 hectares
27. Llyod botanic garden comprises of rosary, aquatic garden, herbarium collections, cacti, succulents, rock garden, bulbous plant collections
28. NBRI is popularly known as Sikander Bagh
29. NBRI garden was laid out by Nawab Saadat Ali Khan
30. Total area of NBRI garden is 27 hectares
31. Rashtrapati Bhavan Garden is located at New Delhi
32. The style of garden laid out in Rashtrapati Bhavan Garden is Mughal
33. Rashtrapati Bhavan Garden is opened for public only during winter months (Jan 15 – Feb 15)
34. Buddha Jayanti Park is laid out at New Delhi
35. Buddha Jayanti Park was established to commemorate 2500th birth anniversary of Lord Buddha
36. Total area of Buddha Jayanti Park is 30 hectares

37. One portion of Buddha Jayanti Park comprises of Japanese style of garden
38. The first Japanese style garden was laid out in Roshnara Park, New Delhi
39. A Japanese landscape architect who gave idea to establish Japanese garden in India is professor K. Mori during 1958
40. Botanical garden, Forest Research Institute, Dehradun occupies an area of 10 hectares
41. Botanical garden, Forest Research Institute, Dehradun maintaining a large collection of terrestrial and epiphytic orchids, ferns and other shade loving plants
42. Descending terraces are the common feature of Mughal type garden
43. A spectacular feature of Mughal garden is chenar tree
44. Garden of Eden is considered as "Genesis of gardening"
45. The first garden developed by Babur after defeating Ibrahim Lodhi is at Paniput
46. Rambagh garden at Agra was developed by Babur
47. The first Mughal emperor who entered Kashmir is Akbar
48. Akbar developed the first garden at Fatehpur Sikri
49. Nasim bagh near Dal lake was developed by Akbar
50. Gardens developed by Jehangir are
    - Gardens at Shalimar
    - Achhabal garden
    - Verinag in Kashmir
    - Tomb garden Itimad-ud-Daulah in Agra
    - Dilkusha garden in Lahore
51. Gardens developed by Shahjahan are
    - Red Fort of New Delhi
    - Taj Mahal of Agra
    - Chasma Shahi of Kashmir
    - Shalimar garden in Lahore

52. Nawab Wajid Ali Shah developed garden at NBRI, Lucknow
53. Sayaji park is located in Baroda (Gujarat)
54. Sayaji park was established during 1879
55. Total area of Sayaji park is 40 hectares
56. A rare branching palm present in Sayaji park is *Hyphaene thebaica*
57. The Baroda museum is opened in Sayaji park during 1894
58. Mandor garden is located at Jodhpur (Rajasthan)
59. A garden in which tea stall is situated at higher terrace is Mandor garden
60. Area of Chandigarh rose garden is 15 hectares
61. Moonlight garden in the Chandigarh rose garden comprises of pure white scented roses
62. Pinjore garden is located in Haryana
63. Number of terraces found in Pinjore garden is 6
64. Total area of Pinjore garden is 25 hectares
65. Pinjore garden was laid out by Fidai Khan
66. A garden which is located at the foothills of Himalayas at an altitude of 600 m is Pinjore garden
67. Three magnificent buildings located in Pinjore garden are Shish mahal, the Rang mahal and the Jal mahal
68. Baradari garden is located at Patiala
69. Baradari garden of Patiala was developed by Maharaja Bhupinder Singh and his son Yadvender Singh of Patiala
70. Plants introduced during Mughal period are chenar, weeping willow, rose, nargis, daffodil, llies, tulip, iris, tuberose, carnation, few annuals
71. Moorish garden is located at Spain
72. French garden designer is Le Notre
73. Garden of Versailles was created by Le Notre
74. First botanical garden was developed at Padua near Venice in Italy during 1543
75. Trees which are associated with Buddh are Ashoka, Sal and Palash

76. Garden at Amritsar was developed by Maharaja Ranjit Singh
77. The plant which is mentioned in the play 'Shakuntala' is MadhviNurjahan who discovered otto of rose while taking bath
78. Ryoanji temple garden in Japan was constructed during Kamakura period
79. Buddha temple in Japan was constructed during Ashoka period
80. A garden in India which is connected with Dal lake by a 1.6 km canal is Shalimar garden
81. Number of terraces found in Shalimar garden is 3
82. First terrace of Shalimar garden comprises of baradari and diwan-e-am
83. Second terrace of Shalimar garden contains Diwan-e-khas
84. Third terrace of Shalimar garden contains a magnificient black stone pavilion
85. Nishat Bagh garden was constructed by Asaf Jah
86. Asaf Jah is brother of Nur Jahan
87. Nishat Bagh garden is called as pleasure garden
88. Number of terraces found in Nishat Bagh garden is 12
89. 12 terraces of Nishat Bagh garden represents the sign of Zodiac
90. Chasma-e-shahi garden is located near Srinagar
91. Chasma-e-shahi garden was constructed by Ali Mardan Khan
92. Number of terraces found in Chasma-e-shahi garden is 3
93. Achabal garden was built by Nur Jahan
94. Main feature of Achabal garden is six vertical waterfalls
95. Bijbehara garden was constructed by Dara Shikoh
96. Bijbehara garden is loctaed near Srinagar
97. Dara Shikoh is the elder son of Shah Jahan

II. Fill in the blanks

1. Botanic garden, Coimbatore was established during.................
2. Sims park, Coonoor was established during.........................

3. The Indian botanic garden, Kolkata was established during ....
4. Llyod botanic garden was established during....................
5. Sayaji park was established during....................
6. The Baroda museum is opened in Sayaji park during ....
7. Number of terraces found in Pinjore garden is....................
8. Lal bagh garden was developed during....................
9. First botanical garden was developed in Venice in Italy during ....
10. Number of terraces found in Shalimar garden is....................
11. Number of terraces found in Chasma-e-shahi garden is ....
12. Number of terraces found in Nishat Bagh garden is....................
13. Number of terraces found in Pinjore garden is....................
14. Area of Chandigarh rose garden is....................
15. Total area of Pinjore garden is....................
16. Total area of Sayaji park is....................
17. Total area of NBRI garden, Lucknow is....................
18. Total area of Llyod botanic garden is....................
19. Total area of Indian botanic garden is....................
20. Total area of Ooty botanic garden is....................

III. Choose the best answer

1. Pinjore garden is laid out by....................

   Asaf Jah Fidai Khan Nur Jahan

2. Baradari garden is located in....................

   Patiala Calcutta Bangalore

3. French garden designer is....................

   Hyder Ali Fidai Khan Le Notre

4. First botanical garden was developed at Padua near Venice in ....

   America Japan Italy

5. Ryoanji temple garden is located in.....................

   Italy Japan America

6. Nishat Bagh garden was constructed by.....................

   Asaf Jah Fidai Khan Nur Jahan

7. Achabal garden was built by.....................

   Dara Shikoh Fidai Khan Nur Jahan

8. Bijbehara garden was constructed by ....

   Fidai Khan Dara Shikoh Nur Jahan

9. Nishat Bagh garden is called as.....................

   Genesis of gardening Pleasure garden Sand garden

10. NBRI is popularly known as.....................

    Rambagh Nishat Bagh Sikander Bagh

IV. Match the following

1.	Red Fort of New Delhi	:	Babur
2.	Achhabal garden	:	New Delhi
3.	Fatehpur Sikri	:	Nawab Saadat Ali Khan
4.	Rambagh garden Babur	:	Shahjahan
5.	Buddha Jayanti Park	:	Jehangir
6.	NBRI garden	:	Akbar
7.	Chasma-e-shahi garden	:	Kodaikanal
8.	Indian botanic garden	:	Coonoor
9.	Llyod botanic garden	:	Haryana
10.	The Bryant park	:	Ali Mardan Khan
11.	Sims park	:	Sibpur, Calcutta
12.	Pinjore garden	:	Darjeeling
13.	Brindavan garden	:	Genesis of gardening
14.	Lalbagh garden	:	Mysore
15.	Garden of Eden	:	Bangalore

Answers

II. Fill in the blanks

1. 1908
2. 1874
3. 1787
4. 1878
5. 1879
6. 1894
7. 6
8. 1760
9. 1543
10. 3
11. 3
12. 12
13. 6
14. 15 hectares
15. 25 hectares
16. 40 hectares
17. 27 hectares
18. 24 hectares
19. 150 hectares
20. 20 hectare

III. Choose the best answer

1. Fidai Khan
2. Patiala
3. Le Notre
4. Italy
5. Japan
6. Asaf Jah
7. Nur Jahan
8. Dara Shikoh
9. Pleasure garden
10. Sikander Bagh

IV. Match the following

1. Shahjahan
2. Jehangir
3. Akbar
4. Babur
5. New Delhi
6. Nawab Saadat Ali Khan
7. Ali Mardan Khan
8. Sibpur, Calcutta
9. Darjeeling
10. Kodaikanal
11. Coonoor
12. Haryana
13. Mysore
14. Bangalore
15. Genesis of gardening

Chapter **3**

Styles of Gardening

I. General information

1. Formal type of garden is laid out in a **symmetrical or a geometrical pattern**
2. Features of a formal garden are **trimmed formal hedges, cypress, ashoka trees**
3. Informal type of garden in which all plants and features are arranged **in a natural way**
4. Wild garden was laid out by **William Robinson**
5. Examples for formal type of gardens are **Mughal, Persian, Indian and French style**
6. Examples for informal type of gardens are **English and Japanese garden**
7. Main features of English garden are **lawn, herbaceous border and rockery**
8. Main features of Italian garden are **massive flight of marble stairs, decorative urns, fountains**
9. Mughal type of garden is a replica of **ancient Persian garden**
10. French style of garden was designed by **Le Notre**
11. Le Notre style is considered as an evolution and mastery of the art of **formal garden**
12. Persian and Japanese style of garden was made based on the idea of **heaven**
13. Main features of Persian garden are **flowing water, crafted materials (masonry, carved and pieced marble stones, highly polished stones), terraces on the hill slopes**

14. In Persian garden, fruit trees represent the symbol of **life**
15. In Persian and Mughal garden, cypress represent the symbol of **death and eternity**
16. 'Nahars' in Persian garden represents **flowing canals**
17. The first Mughal emperor who introduced Mughal garden style in India is **Babar**
18. Main features of Mughal garden are **walls, gates, terrace, nahars, baradari, tomb or a mosque and trees**
19. Shape of Mughal garden is **rectangular or square**
20. Selection of site for Mughal garden is on a **hill slope**
21. Mughal garden gate is studded with **bold iron nails and pointed iron spikes**
22. Number of terraces found in Mughal garden is **7**
23. Seven terraces of Mughal garden represents **seven planets**
24. Water canals in Mughal garden comprises of fountains and are paved with **tiles or marble of blue colour**
25. Baradari in Mughal garden comprises of **12 doors**
26. Fruit trees in Mughal garden symbolizes **life and youth**
27. Flowers grown in Mughal garden are **rose, jasmine, carnation, hollyhock and delphinium**
28. Three major elements of Japanese garden are **stone, water and plants**
29. Important teaching of Japanese garden is **unless a garden has an air of piece it's not worth a place visiting**
30. Types of Japanese garden are **hill, flat, tea, passage and sand garden**
31. The main feature of hill garden is **island**
32. In Japanese, hill garden is named as **Tsukiyama-niwa or Tsukiyama-sansui**
33. Hill garden comprises of **stones, pine trees, untrimmed stepping stones over the walks**
34. Island in hill garden is decorated with **worshipping stone**
35. In Japanese garden, worshipping stone is named as **raithai-seki**
36. In Japanese, flat gardens are called as **hira-niwa**

37. Tea garden is divided into **outer garden and inner garden**
38. Outer garden of tea garden is named as **soto-roji**
39. Inner garden of tea garden is named as **uchi-roji**
40. Gate of tea garden is made up of **bamboo**
41. Fence of tea garden should be **rustic**
42. The entire path leading to tea house is **paved with stones or studded with stepping-stones**
43. Number of persons who can accommodate in a tea house is **five**
44. Main feature at the entrance of tea house is **water basin or a well or both**
45. Inner garden of tea garden contains **tea house**
46. Outer garden of tea garden comprises of **simple planting and stone groupings**
47. Low-door of the tea house simulates **respect and humility**
48. In Japanese, passage garden is named as **Roji-niwa**
49. In Japanese, sand garden is named as **Ryoanji**
50. The most famous sand garden is located in **Kyoto**
51. Main feature of sand garden is an **arrangement of vertical and prostrate stones in groups of 2 or 3 or to fill in the gap between the stones with fine white gravel**
52. Features of Japanese garden are ponds, streams, waterfalls, fountains, wells, islands, bridges, water basins, stone lanterns, stones, pagodas, fences, gates, vegetation and flowering plants
53. Four islands of Japanese garden are **Elysian Isle, Wind-swept Isle, Master's Isle, Guest's Isle**
54. Japanese name for following islands of Japanese garden are

Elysian Isle	Horai-jima
Wind-swept Isle	Fukiage-jima
Master's Isle	Shujin-to
Guest's Isle	Kiakujin-to

55. Pagodas in Japanese garden represents **stone tower**
56. Fences used in Japanese garden are **two types**

57. A type of fence used in Japanese garden for the purpose of screening is **sleeve fence**

58. Types of trees used in Japanese garden are **evergreen, deciduous and flowering**

59. Evergreen trees used in Japanese garden are **pines, different species of Abies, *Cryptomeria japonica, Podocarus macrophylla, Juniperus chinensis***

60. Deciduous trees used in Japanese garden are **maples, poplars, mulberry, *Salix babylonica***

61. Flowering trees used in Japanese garden are **prunus species, *Magnolia grandiflora***

62. Japanese never train their trees or shrubs in **topiary form**

63. Flowering plants grown in Japanese garden are **chrysanthemum, asters, carnation, iris, lotus, peonies and orchids**

64. Climbers used in Japanese garden are ***Clematis, Lonicera japonica, Ipomea hederacea, Ipomea purpurea, Trachelospermum jasminoides, Wisteria sinensis***

II. Choose the best answer

1. In Persian garden, fruit trees represent the symbol of ....

   Life Death Eternity

2. Number of terraces found in Mughal garden is ....

   5 6 7

3. Baradari in Mughal garden comprises of............ number of doors

   10 11 12

4. Gate of tea garden is made up of ....

   Iron Bamboo Steel

5. Flowering tree used in Japanese garden is..................

   Magnolia grandiflora *Cassia fistula* *Tecoma stans*

III. Fill in the blanks

1. In Japanese garden, worshipping stone is named as..................
2. The most famous sand garden is located at ....

3. Pagodas in Japanese garden represents ....
4. Japanese never train their trees or shrubs for..................
5. French style of garden was designed by ....

IV. Match the following

1.	Outer garden of tea garden	: Uchi-roji
2.	Inner garden of tea garden	: Soto-roji
3.	Sand garden	: Death and eternity
4.	Hill garden	: Island
5.	Cypress	: Ryoanji

Answers

II. Choose the best answer

1. Life
2. 7
3. 12
4. Bamboo
5. *Magnolia grandiflora*

III. Fill in the blanks

1. Raithai-seki
2. Kyoto
3. Stone tower
4. Topiary
5. Le notre

IV. Match the following

1. Soto-roji
2. Uchi-roji
3. Ryoanji
4. Island

Death and eternity

Chapter 4

Trees

I. General information

1. Common names of trees

- *Acacia nilotica* – Black babool, Indian gum-Arabic tree
- *Acacia auriculiformis* – Golden shower
- *Adansonia digitata* – Baobab, Monkey bread
- *Adenanthera pavonina* – Red sandal wood, bead tree, red wood
- *Ailanthus excelsa* – tree of heaven, paradise tree
- *Albizia julibrissin* – Pink siris, silk tree
- *Albizia lebbek* – Siris, black siris
- *Alstonia scholaris*- Devil's tree, Dita bark tree
- *Amherstia nobilis* – Tree of heaven, Queen of the flowering trees
- *Andira inermis*- Cabbage tree, Almendro
- *Anthocephalus chinensis*- Kadamba
- *Araucaria bidwillii* – Bunya bunya, Monkey's puzzle
- *Araucaria cookii* – New Caledonia pine, Christmas tree
- *Araucaria cunninghamii* – Moreton Bay pine, Hoop pine
- *Araucaria heterophylla* – Norfolk Island pine
- *Artocarpus altilis* – Bread fruit
- *Azadirachta indica* – Neem tree, Margosa tree
- *Barringtonia acutangula* – Indian oak, Small Indian Oak
- *Barringtonia asiatica*- Fish poison tree

- *Barringtonia racemosa* – Indian oak
- *Bauhinia blakeana* – Hongkong orchid tree
- *Bauhinia purpurea* – Camel's foot tree, Mountain ebony
- *Bauhinia variegata*- Purple orchid tree
- *Bixa orellana* – Annatto dye tree
- *Bombax ceiba*- Red silk- cotton tree
- *Bombax ellipticum* – shaving – bush tree
- *Bombax ellipticum* 'Album' – white shaving bush tree
- *Brownea coccinea* – West Indian Mountain rose, Scarlet flame bean
- *Butea monosperma*- flame of the forest, parrot tree
- *Caesalpinia coriaria* – American sumach, Divi-divi
- *Callistemon citrinus*- Red bottle- brush
- *Calophyllum inophyllum* – Alexandrian laurel dilo oil tree, Punnai
- *Cananga odorata*- ylang- ylang
- *Cassia fistula*- Indian laburnum, Golden shower
- *Cassia grandis*- Horse cassia, Pink shower
- *Cassia javanica*- java cassia
- *Cassia marginata*- red cassia
- *Cassia moschata*- bronze – shower tree
- *Cassia nodosa*- pink cassia, pink mohur
- *Cassia renigera*- Burmese pink cassia
- *Cassia siamea*- kassod tree
- *Castanospermum australe* – black bean
- *Casuarina equisetifolia* – Beef wood, Casuarina, Australian oak
- *Cedrela toona* – Toon, Indian mahogany
- *Ceiba pentandra*- white cotton tree, kapok tree
- *Chorisia speciosa* – Floss silk tree, Mexican silk cotton tree
- *Cochlospermum religiosum* – yellow silk cotton, Buttercup tree
- *Colvillea racemosa* – Colville's glory
- *Cordia sebestena* – Scarlet cordia, aloe wood

- *Couroupita guianensis*- cannon ball tree
- *Crateva religiosa* – Bengal quince, Caper tree
- *Cupressus sempervirens* – Italian cypress
- *Cycas circinalis* – Sissoo, South Indian red wood
- *Delonix regia* – Peacock flower, Flamboyant, Flame tree, Gulmohar
- *Dillenia indica* – Elephant apple
- *Erythrina variegata* – Indian coral tree
- *Eucalyptus variegata*- Lemon-scented eucalyptus
- *Ficus benghalensis* – Banyan tree
- *Ficus benjamina* – Java fig, Java willow
- *Ficus elastica* – Rubber tree, Indian caoutchou tree
- *Ficus religiosa* – Pipal tree
- *Filicium decipiens* – Fern tree, Fern-leaf tree
- *Firmiana colorata* – Bonfire tree
- *Gliricidia sepium*- Mata- raton
- *Grevillea robusta* – Australian silk oak, Silver oak
- *Guaiacumofficinale*- Lignum- vitae, tree of life
- *Gustavia angusta* – Stink wood
- *Haematoxylum campechianum* – Logwod, Campeachy tree
- *Holarrhena antidysenterica*- Easter tree, ivory tree
- *Inga dulcis*- Madras thorn, Manila tamarind
- *Ixora parviflora* – Torch tree
- *Jacaranda mimosaefolia*- Mimosa-leaved Jacaranda
- *Kigelia pinnata*- Cucumber tree, Sausage tree
- *Kleinhovia hospita*- Tree antigonon
- *Lagerstroemia speciosa*- Giant crape myrtle, Pride of India
- *Magnolia grandiflora*- Lily tree, Laurel magnolia
- *Melaleuca leucadendron*- White bottle brush, Cajeput oil tree
- *Melia azedarach*- Persian lilac, bead tree, Indian lilac

- *Mesua ferrea*- Iron-wood tree
- *Michelia champaca*- Golden champa, yellow chamba
- *Millettia peguensis*- Jewel on a string
- *Millingtonia hortensis*- Indian cork tree, tree jasmine
- *Mimusops elengi*- Indian medlar
- *Monodora grandiflora*- African orchid
- *Muntingia calabura*- Chinese cherry/ Singapore cherry
- *Nyctanthes arbor-tristis*- Coral jasmine, night jasmine, pavala malli
- *Parmentiera cereifera*- candle tree
- *Peltophorum pterocarpum*- Rusty shield bearer, Braziletto wood, copperpod tree
- *Pinus longifolia*- long leaved pine, kumaon pine
- *Plumeria* spp.- Frangipani, Pagoda tree
- *Polyalthia longifolia*- Mast tree, Indian fir
- *Pongamia pinnata*- Indian beech, Poonga oil plant
- *Prosopis juliflora*- Algaroba bean, mesquite
- *Pterocarpus indicus*- Burmese rose-wood, Padauk
- *Pterospermum acerifolium*- Muchkunda, Mapel twist
- *Putranjiva roxburghii*- Child life tree, lucky bean tree
- *Samanea saman*- Rain tree, Monkey pod
- *Saraca asoca*- Asoka tree
- *Schleichera oleosa*- Lac tree, Ceylon oak
- *Sesbania grandiflora*- Swamp pea
- *Spathodia campanulata*- Scarlet bell tree, African tulip tree
- *Sterculia foetida*- Dung tree
- *Swietenia macrophylla*- Large-leaved mahogany, bastard mahogany
- *Swietenia mahagoni* – Mahogany, Spanish mahogany
- *Tabebuia argentea*- Golden bell
- *Tabebuia donnell*-smithii – Gold tree

- *Tabebuia rosea*- Rosy trumpet tree
- *Terminalia arjuna*- White murdah
- *Terminalia catappa*- Indian almond
- *Thespesia populnea*- Indian tulip tree, Portia tree, umbrella tree
- *Thevetia peruviana*- Yellow oleander, trumpet flower

2. Trees suitable for moist areas

- *Alstonia scholaris*
- *Alstonia macrophylla*
- *Amherstia nobilis*
- *Barringtonia acutangula*
- *Barringtonia racemosa*
- *Bauhinia variegata*
- *Bauhinia purpurea*
- *Brownea coccinea*
- *Brownea ariza*
- *Cassia marginata*
- *Cassia javanica*
- *Cassia nodosa*
- *Colvillea racemosa*
- *Delonix regia*
- *Dillenia indica*
- *Guaiacum officinale*
- *Lagerstroemia speciosa*
- *Lagerstroemia thorelli*
- *Lagerstroemia flos reginae*
- *Milletia auriculata*
- *Pongamia pinnata*
- *Poinciana regia*

- *Putranjiva roxburghii*
- *Peltophorum pterocarpum*
- *Samanea saman*
- *Polyalthia longifolia*
- *Polyalthia pendula*
- *Saraca asoca*
- *Terminalia arjuna*
- *Tamarindus indica*
- *Solanum wrightii*

3. Trees suitable for marshy areas

- *Barringtonia acutangula*
- *Barringtonia racemosa*
- *Eucalyptus rostrata*
- *Hibiscus tiliaceus*
- *Salix babylonica*
- *Salix tetrasperma*
- *Tamarix articulata*

4. Nectar yielding trees

- *Bauhinia purpurea*
- *Callistemon lanceolatum*
- *Aesculus indica*
- *Eugenia jambolana*
- *Cedrela tuna*
- *Sapium sebiferum*
- *Bauhinia varvigata*
- *Azadirachta indica*
- *Dalbergia sissoo*
- *Crataeva religiosa*

5. Trees suitable for arid & rocky regions

Acacia arabica
Acacia leucophloea
Acacia auriculiformis
Acacia nilotica
Ailanthus excelsa
Albizia lebbek
Albizia procera
Azadirachta indica
Beaumontia grandiflora
Bombax ceiba
Butea monosperma
Callistemon citrinus
Cassia fistula
Casuarina equisetifolia
Caesalpinia coriaria
Ceiba pentandra
Cochlospermum religiosum
Erythrina variegata
Erythroxylon monogynum
Eucalyptus citriodora
Euphorbia antiquorum
Eucalyptus maculata
Euphorbia tirucalli
Ficus tomentosa
Grevillea robusta
Hardwickia binata
Heterophragma adenophyllum
Melia azadirach
Millingtonia hortensis
Morus indica
Parkinsonia aculeata
Pongamia pinnata
Polyalthia longifolia
Plumeria species
Prosopis juliflora
Tamarindus indica
Thevetia nereifolia
Santalum album
Salvadora persica
Sapindus emarginatus
Sterculia foetida
Syzygium cumini
Sterculia villosa

6. Trees suitable for saline soils

- *Acacia auriculiformis*
- *Azadirachta indica*
- *Butea monosperma*
- *Butea frondosa*
- *Bassia latifolia*
- *Casuarina equisetifolia*
- *Cerbera manghas*
- *Cochlospermum religiosum*

- *Eucalyptus citriodora*
- *Melaleuca leucadendran*
- *Phoenix dactylifera*
- *Phyllanthus emblica*
- *Heritiera macrophylla*
- *Inga dulcis*
- *Thespesia populnea*
- *Tamarindus indica*
- *Tamarix articulata*
- *Terminalia arjuna*
- *Thespesia populnea*

7. Trees which are not grazed by cattle

- *Cassia fistula*
- *Cassia nodosa*
- *Cassia javanica*
- *Cassia siamea*
- *Cassia marginata*
- *Cassia renigera* Holarrhena
- *Cassia multijuga* Antidysenterica

8. Tree species which are propagated by seeds

Acacia spp.
Albizia spp.
Alstonia scholaris
Bauhinia spp.
Araucaria spp.
Cassia spp.
Delonix regia
Dalbergia sissoo
Anthocephalus chinensis
Azadirachta indica
Bombax ceiba
Calophyllum inophyllum
Ceiba pentandra
Castanospermum australe
Cochlospermum religiosum
Colvillea racemosa

Eucalyptus spp.
Jacaranda spp.
Lagerstroemia speciosa
Michelia champaca
Milletia spp.
Peltophorum spp.
Polyalthia spp.
Pterospermum spp.
Saraca spp.
Sterculia spp.
Swietania mahagoni
Terminalia spp.
Couroupita guianensis
Erythrina variegata
Guaicum officinale
Gustavia angusta
Heterophragma adenophyllum
Kigelia pinnata
Kleinhovia hospita
Mimusops elengi
Millingtonia hortensis
Putranjiva roxburghii
Thespesia populnea

9. Tree species which are propagated by cutting & layering

Anthocephalus chinensis
Barringtonia racemosa
Bauhinia blackeana
Bauhinia variegata
Bauhinia purpurea
Bombax ceiba
Brownea arixa
Brownea grandiceps
Butea monosperma
Elaeodendron glabrum
Eriolaena wallichi
Ficus cannonii
Ficus krishnae
Muntingia calabura
Napoleona imperialis
Erythrina variegata
Podocarpus macrophylla
Mimusops elengi
Sterculia spp
Pterospermum semisagitattum
Pterygota alata
Saraca indica
Callistemon lanceolatus
Calophyllum inophyllum
Ceiba pentandra
Citharexyllum quadrangularae
Crataeva religiosa
Erythrina cristagalli
Ficus benjamina
Ficus elastica
Gustavia angusta
Lagerstroemia speciosa
Millietia ovalifolia
Parmentiera cereifera
Peltophorum pterocarpum
Gliricidia sepium
Terminalia spp
Thespesia populnea

10. Tree species which are propagated through micro propagation

- *Grevillea robusta*
- *Eucalyptus* spp.
- *Callistemon viminalis*
- *Albizia lebbeck*

11. Tree species which produce scented flowers

Albizia lebbeck
Alstonia scholaris
Anthocephalus indicus
Artabotrys hexapetalus
Calophyllum inophyllum
Couroupita guianensis
Cananga odorata
Citharexylum guianensis
Nyctanthes arbortristis
Pandanus odoratissimus
Pterospermum acerifolium
Fragraea obovata
Magnolia grandiflora
Mesua ferrea
Michelia champaca
Millingtonia hortensis
Mimusops elengi
Murraya exotica
Poeciloeneuron indicum
Polyalthia fragrans
Saraca indica

12. Tree species which are suitable for hedges, screens & barriers

Bougainvillea spectabilis
Caesalpinia pulcherrima
Caesalpinia coriaria
Casuarina equisetifolia
Dodonaea viscosa
Duranta repens
Euphorbia antiquorum
Grevillea robusta
Lantana camara
Murraya exotica
Pisonia alba
Polyalthia longifolia
Prosopsis juliflora
Putranjiva roxburghii
Tabebuia rosea
Tamarindus indica
Thevetia nereifolia
Triphasia trifolia

13. Trees which are used as specimen tree

Bauhinia tomentosa
Bixa orellana
Mussaenda philippica
Nyctanthes arbor-tristis

Callistemon citrinus
Delonix regia
Ixora arborea
Polyalthia longifolia
Thuja orientalis

14. Trees suitable for avenues, streets & woodlands

Adansonia digitata
Artocarpus heterophyllus
Azadirachta indica
Caryota urens
Cochlospermum religiosum
Ceiba pentandra
Ficus benghalensis
Ficus tomentosa
Ficus tsiela
Polyalthia longifolia
Grevillea robusta
Madhuca longifolia
Muntingia calabura
Oreodoxa regia
Parkia pedunculata
Peltophorum pterocarpum
Phoenix sylvestris
Pongamia pinnata
Tamarindus indica
Terminalia arjuna
Terminalia paniculata

15. Trees unsuitable for roadside avenues

Brittle trees

Eucalyptus spp.
Millingtonia hortensis
Eugenia jambolana
Albizzia lebbek
Ficus glomerata

Thorny trees

Acacia arabica
Acacia modesta
Zizyphus jujuba
Cassia siamea

16. Ornamental trees suitable for town roads

Foliage trees for outer avenues

Averrhoea carambola
Callistemon lanceolatus
Anthocephalus cadamba
Eugenia operculata
Polyalthia longifolia

Foliage trees for inner avenues

Cassia fistula
Bauhinia purpurea
Bauhinia variegata
Colvillea racemosa
Peltophorum ferrugineum

Putranjiva roxburghii
Sterculia alata
Pithecolobium dulci
Melia azadirachta
Tamarindus indica
Grevillea robusta
Spathodea nilotica
Jacaranda mimosaefolia
Poinciana regia
Lagerstroemia flosreginae
Lagerstroemia thorelli
Gliricidia maculata

17. Dwarf ornamental trees suitable for small compounds

Acacia auriculiformis
Alangium lamarckii
Bauhinia purpurea
Bauhinia variegata
Butea frondosa
Cassia fistula
Cassia javanica
Cassia marginata
Cochlospermum gossypium
Cordia sebestena
Crataeva religiosa
Erythrina blakei
Erythrina cristagalli
Gliricidia maculata
Guaiacum officinale
Holarrhena antidysenterica
Jacaranda minosaefolia
Brownea coccinea
Brownea ariza
Kleinhovia hospita
Lagerstroemia thorelli
Mesua ferrea
Milletia auriculata
Plumeria rubra and *Plumeria alba*
Pongamia glabra
Saraca indica
Solanum wrightii
Spathodea nilotica
Sterculia colorata
Tecomella undulata
Thespesia populnea

18. Trees suitable for sodic clay soils

Scientific Name	Common Name	Landscape features
Acacia arabica	Babul	It can stand in stagnant water as well as in dry places
Cassia marginata	Red konnai	A native of this soil type
Cordia sebestina	-	A flowering tree
Dalbergia sissoo	Sisoo	Requires soil moisture for good growth
Delonix elata	Vada Narayanam	A drought-hardy tree

Madhuca longifolia	Mahwa	A spreading tree
Malpighia glabra	West-Indian cherry	Survives drought, but summer irrigation is essential for fruiting
Manilkara hexandra	Manilkara	Grows in moist places
Peltophorum pterocarpum	Copper-pod	Tolerant to drought
Samanea saman	Rain tree	Tolerant to alternating wet and dry conditions
Tabebuia rosea	-	Of fast growth, where water is available
Terminalia arjuna	Arjuna tree	Loves a plentiful supply of water
Thespesia populnea	Portia tree	Both in dry and moist situations

19. For beaches and neighbouring areas

Scientific Name	Common Name	Landscaping features
Alstonia scholaris	Ezhila pala	An upright tree, grows even in poor sand
Beaumontia grandiflora	Rubber vine	A climbing shrub, very vigorous in growth
Calophyllum inophyllum	Poon	A medium-sized tree
Casuarina equisetifolia	She-oak	A useful tree for wind-break screens and soil binding
Guaiacum officinale	W.I. tree of life	A tree of strong framework, resistant to wind
Madhuca longifolia	Mahwa	A spreading tree
Manilkara hexandra	Manilkara	A small tree, grows even in poorly drained situations
Mimusops elengi	Elengi	A medium spreading tree
Plumeria rubra	Frangipani	Very tolerant to salt spray
Pongamia pinnata	Pungam	A medium-sized shade and flowering tree
Samanea saman	Rain tree	A spreading tree of fast growth
Spathodea campanulata	Indian tulip	An upright tree, flowering in profusion
Tabebuia rosea	-	A flowering tree of fast growth
Thespesia populnea	Portia tree	Cultivated for shade and flowers

20. To plant as free-standing trees

Scientific Name	Common Name	Landscape features
Azadirachta indica	Neem	A tree of sentimental value
Calophyllum inophyllum	Poon	Medium evergreen, lovely white scented flowers
Ceiba pentandra	White silk cotton	Large, conical tree, flowers attractive
Citharexylum spinosum	Fiddle wood	Small to medium upright tree, fast in growth, scented flowers
Cochlospermum religiosum	Yellow silk cotton	Deep shade, yellow flowers
Cordia sebestina		Small tree, deep shade, evergreen, orange flowers
Ficus benjamina	Java fig	Medium spreading, weeping in habit
Gliricidia maculata	Madera	Medium, spreading, light rose-tinged flowers
Guaiacum officinale	W.I. tree of life	A small tree, resistant to wind of deep shade, blue flowers
Mimusops elengi	Elengi	Medium, evergreen, spreading, scented flowers
Spathodea campanulata	Indian Tulip	Medium, orange-red flowers, appearing in two or three seasons
Tecoma stans	-	Large woody shrub, flowers bright yellow

21. Endemic and indigenous flowering trees

Scientific Name	Common Name	Flower colour and season
Bauhinia purpurea	-	Purple ; January-February, September
Butea monosperma	Flame of the forest	Red ; April-May
Cassia fistula	Golden shower	Golden yellow ; April-May
Cassia marginata	Red Konnai	Red ; April-June
Delonix elata	Vada Narayanam	Yellowish- White ; February-July
Lagerstoremia speciosa	Pride of India	Mauve purple ; May-June
Michelia champaca	Champakam	Light yellow; April-May

Mimusops elengi	Elengi	Grey-white; April-May
Nyctanthes arbor-tristis	Parijatham	White with orange tube; February-May; October-December
Pterospermum acerifolium	Ven champakam	White ; March-May, August-November
Polyalthia fragrans	-	Greenish white; May-June

22. Accent trees

Scientific Name	Common Name	Landscape features
Bauhinia purpurea	Mandaram	A small tree, purple flowers
Cassia biflorus	-	A dense shrub, bright yellow flowers
C. fistula	Golden shower	A small to medium tree, flowers in golden spangles
C. marginata	Red 'Konna'	Small to medium; brick red flowers
Cochlospermum religiosum	Yellow silk cotton	A small tree, showy yellow flowers
Delonix regia	Gulmohar	A picturesque tree, with deep orange flowers
Lagerstoremia speciosa	Pride of India	A medium-sized tree, with mauve purple flowers
Saraca indica	Asoka	Beautiful foliage, orange flowers in summer

23. Specimen trees

Scientific Name	Common Name	Landscape features
Adenium obesum	Desert rose	Trunk caudatus, showy rose coloured flowers
Bauhinia tomentosa	Mandaram	Vigorous shrub, yellow flowers
Bixa orelana	Annato	A large shrub, attractive flowers and spiny pods
Callistemon citrinus	Bottle-brush	Small tree, weeping, fine in texture
Cycas circinalis	Cycad growth	Palm like in habit, slow in

Cycas revoluta	Cycad	Similar to the above species, with shorter leaf
Delonix regia	Gulmohar	A picturesque tree with deep orange flowers in summer
Polyalthia longifolia (grafted)	Nettilingam	Tall, weeping
Tamarix gallica	Tamarisk	A small tree, the leaf-like Cladode fine textured
Thuja orientalis	-	A low conifer, with fine textured foliage

24. Naturalized flowering trees of exotic origin

Scientific Name	Common Name	Flower colour & season
Acacia auriculiformis	Babul	Yellow flowers in February-March and August-September
Cassia javanica	Red Cassia	Flowers red, April-May
Ceiba pentandra	White silk cotton	White, February - March
Citharexylum spinosum	Fiddle wood	A strong, small to medium-sized tree, scented flowers, good even for infertile sand
Colvillea racemosa	Colville's glory	Orange to red, May-June
Couropita guianensis	Nagalingam	March-April, spread over the other parts of the year also
Guaiacum officinale	W.I. tree of life	Blue; April-May ; October - November
Jacaranda mimosifolia	-	Blue; April-May ; October - November
Millingtonia hortensis	Indian cork tree	White; February - June ; October -November
Peltophorum pterocarpum	Copper pod	Yellow ; April-June; spread over the other parts of the year also
Plumeria rubra	Frangipani	White or red; February - June
Samanea saman	Rain tree	Rosy purple; April - May; September - October

25. Selected trees for warm humid and hilly regions

Scientific Name	Common Name	Landscape features
Alstonia scholaris	Ezhilapala	A tall tree, flowering at the beginning of summer
Anthocephalus cadamba	Kadambu	Of fast growth, deep orange flowers
Cassia fistula	Golden shower	Known for its golden spangles of flowers
Citharexylum spinosum	Fiddle wood	A fast growing tree, scented flowers
Delonix regia	Gulmohar	A familiar land-mark now
Guaiacum officinale	W.I. tree of life	A strong, wind resistant flowering tree
Gmelina arborea	Kumil	Bright yellow flowers in summer
Ixora (all species and cultivars)	Thechi	Shrubs, attractive in flower
Mussaenda (all species and cultivars)	Mussanda	Possesses attractive calycine leaves
Saraca indica	Asoka	A small flowering tree
Spathodea campanulata	Indian tulip	A tall flowering tree
Swietenia mahagoni	Mahogani	A strong tree, wind resistant
Tabebuia argentea	-	Yellow flowers

26. Trees for water margins and inundated areas

Scientific Name	Common Name	Remarks
Acacia arbica	Babul	Stands in water for long periods
Butea monosperma	Flame of the forest	Of picturesque growth, red flowers in summer, for margins of inland water sources
Casuarina equisetifolia	She-oak	Useful for inland and maritime planting in inundated areas
Ceiba pentandra	White silk cotton	For streams and river margins
Cycas revoluta	Cycad	For fresh-water pool embankment

Lagerstroemia speciosa	Pride of India	For stream and river margins
Salix tetrasperma	Willow	A strong foliage tree on the margins of stream and fresh-water ponds
Syzygium cumini	Jambolanum	A medium to large-sized tree, useful near large bodies of water
Tamarix gallica	Tamarisk	Tolerates brackish and saline water
Terminalia arjuna	Arjuna	For inundated islands

27. Trees for fruits, edible or otherwise

Scientific Name	Common Name	Remarks
Adansonia digitata	Baobab	Edible large pods
Anacardium occidentale	Cashew	Cashew apple attractively coloured
Annona squamosa	Custard apple	Well known fruit trees, fruitful even in shade of other trees
Ardesia crenata	-	Showy berries in clusters, not edible
Artocarpus heterophyllus	Jackfruit	A well known fruit
Ficus benghalensis	Banyan	A haven for frugivorous birds and animals
Ficus glomerata	Indian fig	Fruits attract birds and small animals, cauliflorous
Ficus macrophylla	-	Cauliflorous, provides food for bats and birds
Ficus religiosa	Peepul	A haven for birds
Malpighia glabra	West Indian cherry	Fruits sub-acid, edible
Mangifera indica	Mango	The mango is too well known for description
Muntingia calabura	Singapore cherry	A tree of dense growth, fruits small, sweet, equally relished by birds
Polyalthia longifolia	Nettulingam	Fruits attract Koel
Triphasia trifolia	Lime-berry	A spiny shrub, fruits sub-acid, edible

28. Cauliflorous trees

Scientific Name	Common Name	Landscape features
Artocarpus heterophyllus	Jack tree	A large fruit tree, evergreen
Averrhoa bilimbi	Bilimbi	A small fruit tree, fruit tart
Couroupita guianensis	Nagalingam	Showy scented flower, fruits of the size of a cannon ball
Crescentia cujete	West Indian calabash tree	A small tree, flowers showy, garlic scented, fruit has a shell
Durio zibethinus	Durian	Edible fruits, spiny, for humid situations
Ficus racemosa	Indian fig	A medium to large tree, greenish yellow fruits, loved by animals and birds
Phyllanthus acidus	Star gooseberry	Tart ridged fruits
Saraca indica	Asoka	A small flowering tree, old trees and cauliflorous

29. Trees for tubs

Scientific Name	Common Name	Landscape features
Bougainvillea spectabilis	-	Blooms freely in tubs
Calophyllum inophyllum	Poon	Attractive foliage, flowers and fruits
Cassia fistula	Golden showers	Free blooming in summer
Ficus benghalensis	Banyan	Train the trees to show off their adventitious roots
Ficus benjamina	Java fig	
Ficus deltoidea	-	
Ficus religiosa	Peepul	
Mimusops elengi	Elengi	Attractive foliage
Murraya exotica	Kamini	A flowering shrub or small tree
Polyalthia longifolia	Nettulingam	Attractive foliage
Pterospermum acerifolium	Ven champakam	Grafted trees flower freely
Syzygium cumini	Jambolanum	Stands heavy pruning
Tabebuia rosea	-	Attractive foliage, free blooming in tubs
Thespesia populnea	Portia tree	Start from limb cuttings which flower in course of time

30. Trees suitable for both sunny and semi-shady condition

Botanical Name	Common name	Family	Flowering season	Flower colour
Brassaia actinophylla	Queensland umbrella tree	Araliaceae	-	Red
Brownea coccinea	West Indian mountain rose	Caesalpiniaceae	Feb-March & also in September	Pink
Cassia siamea	Kassod tree	Caesalpiniaceae	June – Jan Peak in autumn	Yellow
Thevetia peruviana	Yellow oleander	Apocynaceae	Throughout the year	Yellow

31. Trees suitable for semi-shady location

Botanical Name	Common name	Family	Flowering season	Flower colour
Napoleona imperialis	-	Lecythidaceae	Feb - June	Brown, rose & yellowish
Saraca declinata	Red saraca	Caesalpiniaceae	Late Feb - April	-
Saraca thaipingensis	Yellow saraca	Caesalpiniaceae	Jan - Mar	Golden yellow
Cerbera tanghin	-	Apocynaceae	-	White
Mesua ferrea	Iron-wood tree	Guttiferae	Feb - May	White, fragrant
Podocarpus macrophylla	Buddhist pine	Podacarpaceae	-	-

32. Trees suitable for shady location

Botanical Name	Common name	Family	Flowering season	Flower colour
Artocarpus altilis	Bread fruis	Moraceae	-	-
Azadirachta indica	Neem tree	Meliaceae	March-April	White, honey scented
Caesalpinia coriaria	Divi Divi	Caesalpiniaceae	April -July	Greenish yellow (scented)
Castanospermum australe	Moreton Bay chestnut	Caesalpiniaceae	March	Yellow to orange
Eucalyptus calophylla	-	Myrtaceae	-	White (or) cream colour
Ficus elastica	Indian Rubber tree	Moraceae	-	-
Ficus infectoria	-	Moraceae	-	-
Ficus religiosa	Pipal tree	Moraceae	-	-
Ficus rumphii	-	Moraceae	-	-
Klenihovia hospita	Tree antigonon	Sterculiaceae	July - Sep	Rose coloured
Leucaena leucocephala	Horse tamarind	Mimosaceae	Mar – April & again Aug - Nov	White
Swietenia mahagoni	Spanish mahogany	Meliaceae	Summer	Green
Thespesia populnea	Indian tulip tree/ Portia tree	Malvaceae	April - May	Yellow & turn purplish with age

33. Trees having slow growing nature

Botanical Name	Common name	Family	Flowering season	Flower colour
Swietenia mahagoni	Spanish mahogany	Meliaceae	Summer	Green
Araucaria hetrophylla	Norfolk Island pine/ Indoor fir	Araucariaceae	-	-
Barringtonia acutangala	Indian Oak	Lecythidaceae	April - May	Pink to Red
Cassia marginata	Red cassia	Caesalpiniaceae	June -July	Terracotta
Crataeva religiosa	Bengal quince/Caper tree	Capparidaceae	Nov-Dec	White (or) cream coloured
Eucalyptus calophylla	-	Myrtaceae	-	White (or) cream colour
Filicium decipiens	Fern-leaf tree	Sapindaceae	-	white
Guaiacum officinale	Tree of Life	Zygophyllaceae	-	Blue
Haematoxylon campechianum	Logwood/Campeachy tree	Mimosaceae	-	Yellow scented
Juniperus chinensis	Chinese juniper	Cupressaceae	-	-
Melaleuca leucadendran	White bottle brush/ Swamp tea tree	Myrtaceae	Hot & rainy seasons	whitish
Mesua ferrea	Iron-wood tree	Guttiferae	Feb - May	White, fragrant
Peltophorum africanum	-	Caesalpiniaceae	Light yellow	June - Aug
Schleichera oleosa	Lac tree/ Ceylon oak	Sapindaceae	Spring	Yellow
Swietenia mahagoni	Spanish mahogany	Meliaceae	Summer	Green
Araucaria hetrophylla	Norfolk Island pine/ Indoor fir	Araucariaceae	-	-

34. Trees with quick growing nature

Botanical Name	Common name	Family	Flowering season	Flower colour
Cassia multijuga	-	Caesalpiniaceae	Sep - Nov	Yellow
Cochlospermum religiosum	Yellow silk cotton	Cochlospermaceae	Feb - March	Brilliant yellow
Delonix regia	Flame of forest	Caesalpiniaceae	April - Aug	Orange -scarlet
Erythrina variegata	Indian coral tree	Caesalpiniaceae	Feb - March	Red
Eucalyptus citriodora	Lemon scented eucalyptus	Myrtaceae	March - May	white
Ficus elastica	Indian Rubber tree	Moraceae	-	-
Ficus infectoria	-	Moraceae	-	-
Gliricidia sepium	Mata-raton	Caesalpiniaceae	-	Pink, white, purpe
Ipomoea arborescens	-	Convolvulaceae	Summer	White
Lagerstroemia parviflora	-	Lythraceae	Summer	white
Leucaena leucocephala	Horse tamarind	Mimosaceae	Mar – April & again Aug - Nov	White
Milligtonia hortensis	Tree jasmine/Indian cork tree	Bignoniaceae	-	White with greenish tinge, fragrant
Parkia biglandulosa	Ball batmitton tree	Mimosaceae	Nov - Dec	Brown & velvety but becomes white as the flowers open
Pterygota alata	Buddha's coconut	Sterculiaceae	Feb - March	Greenish purple (or) rusty brown outside & red veined within
Sesbania grandiflora	Swamp pea	Caesalpiniaceae	Sep - Nov	White, red, pink

Sterculia villosa	Dung tree	Sterculiaceae	-	Yellow with purple centre
Syzygium cumini	Indian blackberry/ Java plum	Myrtaceae	Mar - June	White, fragrant
Thespesia populnea	Indian tulip tree/ Portia tree	Malvaceae	April - May	Yellow & turn purplish with age
Acacia auriculiformis	Golden shower	Mimosaceae	Sep-Oct	Yellow scented
Adenanthera pavonina	Red sandalwood/ Bead tree	Mimosaceae	End of March – AprilSometimes in Sep-Oct	Yellow & Scented
Ailanthus excelsa	Tree of heaven/ Paradise tree	Simarubaceae	End of cold season	White (or) yellow
Albizzia chinensis	-	Mimosaceae	-	White
Albizzia richardiana	-	Mimosaceae	-	White
Cananga odorata	Ylang-ylang	Annoaceae	May - Oct	Pale – yellow fragrant

35. Trees with evergreen nature

Botanical Name	Common name	Family	Flowering season	Flower colour
Acacia auriculiformis	Golden shower	Mimosaceae	Sep-Oct	Yellow scented
Aleurietes moluccana	Candle nut/Varnish tree	Mimosaceae	March - June	White
Amherstia nobilis	Queen of flowering trees	Caesalpiniaceae	Feb-April	Petals brilliant red, edged with yellow
Andira inermis	Cabbage tree	Caesalpiniaceae	March-April	Purple
Anogeissus acuminate	-	Combrataceae	Feb-March	Yellowish white, fragrant
Baikiaea insignis	-	Caesalpiniaceae	April - June	Snow white with lower petal (or) lip lemon yellowFlower buds are large, dark brown& velvety
Barringtonia asiatica	Fish poison tree	Lecythidaceae	July to Sep	White
Bauhinia purpurea	Camel's foot tree	Caesalpiniaceae	October	-
Bixa orellana	Annatto dye tree	Bixaceae	Sep - Dec	Rose, Purple (or) white
Brassaia actinophylla	Queensland umbrella tree	Araliaceae	-	Red
Brownea coccinea	West Indian mountain rose	Caesalpiniaceae	Feb-March & also in September	Pink
Brownea grandiceps	Rose of venezulea	Caesalpiniaceae	March - May	Reddish purple
Brownea hybrida	-	Caesalpiniaceae	Summer months	Reddish purple
Callistemon citrinus	Red bottle-brush	Myrtaceae	-	Calyx pale greenish, filaments numerous, free, bright red
Calophyllum inophyllum	Dilo oil tree	Guttifeerae	-	Sepals and petals are in white colour, stamens yellow

Canaga odorata	Ylang-ylang	Annoaceae	May - Oct	Pale – yellow fragrant
Cassia nodosa	Pink cassia	Caesalpiniaceae	March - April	Bright pink
Cassia siamea	Kassod tree	Caesalpiniaceae	June – Jan Peak in autumn	Yellow
Cassia spectabilis	-	Caesalpiniaceae	Sep - Dec	Golden yellow
Castanospermum australe	Moreton Bay chestnut	Caesalpiniaceae	March	Yellow to orange
Cedrela microcarpa	-	Meliaceae	-	white
Cedrela toona	Indian mahogany	Meliaceae	Feb - March	White fragrant
Cerbera tanghin	-	Apocynaceae	-	White
Chrysophyllum cainito	Star apple	Sapotaceae	-	Purplish white
Clitoria arborea	-	Caesalpiniaceae	May - June	Shades of violet, centre deeper with red markings
Clusia rosea	Fat pork tree	Guttiferae	Jan - March	White with pink tinge
Cordia sebestena	Aloe wood	Boraginaceae	Throughout the year	Orange scarlet
Cycas circinalis	Fern palm	Cycadaceae	-	-
Dillenia indica	Elephant apple	Dilleniaceae	June - July	White, sweet scented
Diospyros peregrina	River ebony	Ebeneceae	April - May	White, scented
Ficus benjamina	Java fig	Moraceae	-	-
Filicium decipiens	Fern-leaf tree	Sapindaceae	-	white
Guaiacum officinale	Tree of Life	Zygophyllaceae	-	Blue
Heritiera macrophylla	-	Sterculiaceae	Mar - April	Pinkish rose

Heterophragma adenophyllum	-	Bignoniaceae	June - Aug	Corolla is of yellowish brown tube, swollen at the top
Inga dulcis	Manilla tamarind	Euphorbiaceae	Jan - April	-
Ixora parviflora	Torch tree	Rubiaceae	Mar - May	White, rarely pink, scented
Kigelia pinnata	Sausage tree/cucumber tree	Bignoniaceae	Mar - May	Deep chocolate red, evil smelling
Magnolia grandiflora	Lily tree	Magnoliaceae	Summer	White, sweet - scented
Magnolia pterocarpa	-	Magnoliaceae	April - May	White, fragrant
Markhamia platycalyx	-	Bignoniaceae	Feb - Mar	Yellow with red lines
Melaleuca leucadendron	White bottle brush/ Swamp tea tree	Myrtaceae	Hot & rainy seasons	whitish
Mesua ferrea	Iron-wood tree	Guttiferae	Feb - May	White, fragrant
Michelia champaca	Yellow champa	Magnoliaceae	Summer & rains	Yellow (or) dull orange, strongly scented
Mimusops elengi	Indian Medlar	Sapotaceae	Spring & throug-hout warm season	Off – white fragrant
Napoleona imperialis	-	Lecythidaceae	Feb - June	Brown, rose & yellowish
Pachira cyathophora	-	Bombacaceae	Mar - May	Cream - white
Parkia biglandulosa	Ball batmitton tree	Mimosaceae	Nov - Dec	Brown & velvety but becomes white as the flowers open
Parmentiera edulis	Guajilote tree	Bignoniaceae	-	Greenish yellow
Podocarpus elongata	African yellow wood	Podocarpaceae	-	-
Podocarpus macrophylla	Buddhist pine	Podacarpaceae	-	-

Polyalthia coffeoides	-	Annonaceae	-	Greenish yellow
Pterospermum suberifolium	-	Sterculiaceae	Summer	Yellowish white scented
Putranjiva roxburghii	Child-life tree/ Lucky bean tree	Sterculiaceae	Mar - May	Yellow
Santalum album	Sandal tree	Santalaceae	-	Purple
Sapium sebiferum	Chinese tallow tree	Euphorbiaceae	April – June & Oct - Nov	-
Saraca asoca	Asoka tree	Caesalpiniaceae	Mar - May	Orange red
Spathodea campanulata	African tulip tree/ Scarlet bell tree	Bignoniaceae	Jan - Feb	Orange crimson outside, having yellow wrinkled margin
Stereospermum kunthianum	-	Bignoniaceae	April - June	Dark purple changing to pale purple
Swietenia macrophylla	Large leaved mahogany	Meliaceae	-	White
Syzygium cumini	Indian blackberry/ Java plum	Myrtaceae	Mar - June	White, fragrant
Syzygium malaccense	Malay rose-apple	Myrtaceae	Summer & Nov - Dec	Crimson
Thespesia populnea	Indian tulip tree/ Portia tree	Malvaceae	April - May	Yellow & turn purplish with age
Thuja orientalis	Chine thuja	Cupressaceae	-	Bluish green

36. Trees with deciduous nature

Botanical Name	Common name	Family	Flowering season	Flower colour
Adansonia digitata	Monkey bread/Sour gourd	Bombacaceae	June/July	Creamy white, unpleasant smell
Ailanthus excelsa	Tree of heaven/ Paradise tree	Simarubaceae	End of cold season	White (or) yellow
Albizzia lebbek	Black siris/Parrot tree	Mimosaceae	-	Creamy white, fragrant
Anthocephalus chinensis	Kadamba	Rubiaceae	June to August	Orange, scented, small silk
Artocarpus altilis	Bread fruis	Moraceae	-	-
Azadirachta indica	Neem tree	Meliaceae	March-April	White, honey scented
Bauhinia variegata	Purple orchid tree	Caesalpiniaceae	Jan - March	Purple pink
Bombax ceiba	Red silk cotton tree	Bombacaceae	February	Red
Bombax ellipticum	Shaving brush tree	Bombacaceae	-	Purple
Caesalpinia coriaria	Divi Divi	Caesalpiniaceae	April -July	Greenish yellow (scented)
Careya arborea	Wild guava	Lecythidaceae	Early summer	Pink & white (foul smell)
Cassia fistula	Golden shower	Caesalpiniaceae	Early summer to Sep	Yellow
Cassia renigera	Burmese pink cassia	Caesalpiniaceae	April - May	Pink scented
Casuarina equisetifolia	Beefwood/Australian oak	Casuarinaceae	-	Pinkish
Ceiba pentandra	Kapok tree	Bombacaceae	Jan - March	Creamy white (or) light pink
Chorisia speciosa	Floss silk tree	Bombacaceae	October	Pinkish outer surface, lower half of inner surface pinikish-white gradually fading

Citharexylum quadrangulare	Zither-wood	Verbenaceae	March - May	White scented
Cochlospermum religiosum	Yellow silk otton	Cochlospermaceae	Feb - March	Brilliant yellow
Colvillea racemosa	Colville's Glory	Caesalpiniaceae	Sep - Oct	Orange
Couroupita guianensis	Cannon ball tree	Lecythidaceae	Several times in a year	Outside a mixture of white, yellow & inside deep pink (or) crimson
Crataeva religiosa	Bengal quince/Caper tree	Capparidaceae	Nov-Dec	White (or) cream coloured
Dalbergia sissoo	South Indian Rewood	Caesalpiniaceae	-	Yellowish white scented
Delonix regia	Flame of forest	Caesalpiniaceae	April - Aug	Orange -scarlet
Dolichandrone spathacea	-	Bignoniaceae	March - May	white
Erythrina variegata	Indian coral tree	Caesalpiniaceae	Feb - March	Red
Ficus bengalensis	Banyan tree	Moraceae	-	-
Ficus infectoria	-	Moraceae	-	-
Gardenia latifolia	Cape jasmine	Rubiaceae	Summer	White, highly scented
Gliricidia sepium	Mata-raton	Caesalpiniaceae	-	Pink, white, purpe
Holarrhena antidysenterica	Easter tree/Ivory tree	Apocynaceae	-	White (or) creamy scented
Jacaranda mimosaefolia	Mimosa – leaved jacaranda	Bignoniaceae	Mar - June	Bluish purple
Lagerstroemia speciosa	Pride of India	Lythraceae	April - June	Mauve
Melia azedarach	Indian lilac/China berry	Meliaceae	-	Purple in colour, scented
Millettia peguensis	Jewels on a string	Caesalpiniaceae	Mar - April	Red (or) purplish

Oroxylum indicum	Indian trumpet flower	Bignoniaceae	June - July	Lurid - purple
Peltophorum pterocarpum	Rusty shield bearer	Caesalpiniaceae	Mar – June & also in Sep - Nov	Yellow
Polyalthia longifolia	Mast tree/ Indian fir	Annonaceae	Mar - April	Greenish yellow
Pongamia pinnata	Indian beech/Poonga oil plant	Caesalpiniaceae	-	Rose (or) liliac
Prosopis julifora	Algaoba bean/Mesquite	Mimosaceae	April - May	Greenish turning yellow, fragrant
Pterocarpus indicus	Burmese rose-wood/ Padauk	Caesalpiniaceae	-	Golden yellow fragrant
Pterospermum semisagittatum	-	Sterculiaceae	-	White
Pterygota alata	Buddha's coconut	Sterculiaceae	Feb - March	Greenish purple (or) rusty brown outside & red veined within
Salix babylonica	-	Saliaceae	-	-
Samanea saman	Rain tree	Mimosaceae	April - Oct	Yellow
Schleichera oleosa	Lac tree/ Ceylon oak	Sapindaceae	Spring	Yellow
Spathodea campanulata	African tulip tree/Scarlet bell tree	Bignoniaceae	Jan - Feb	Orange crimson outside, having yellow wrinkled margin
Sterculia villosa	Dung tree	Sterculiaceae	-	Yellow with purple centre
Swietenia mahagoni	Spanish mahogany	Meliaceae	Summer	Green
Tabebuia avellanedae	-	Bignoniaceae	Winter	Reddish mauve

Tabebuia chrysantha	-	Bignoniaceae	Feb - Mar	Bright yellow
Tabebuia rosea	Rosy trumpet tree	Bignoniaceae	-	Rosy - pink
Tamarindus indica	Tamarind	Caesalpiniaceae	May - Aug	Red & yellow
Taxodium distichum	Bald cypress/Swamp cypress	Pinaceae	-	Purple
Tectona grandis	Teak	Verbenaceae	July - Aug	White
Terminalia arjuna	White murdah	Combretaceae	April - May	Dull yellow
Terminalia catappa	Indian almond	Combretaceae	Mar - April	White
Tipuana tipu	Pride of Bolivia/Rosewood	Caesalpiniaceae	Several months from summer	Bright golden
Wrightia tomentosa	-	Apocynaceae	April - July	Vermillion

37. Trees having tall growing nature (15 m and above)

Botanical Name	Common name	Family	Flowering season	Flower colour
Ailanthus excelsa	Indian Tree of heaven/ Paradise tree	Simarubaceae	End of cold season	White (or) yellow
Albizzia chinensis	-	Mimosaceae	-	white
Albizia procera	White siris	Mimosaceae	May - Sep	Greenish white
Aleurietes moluccana	Candle nut/Varnish tree	Mimosaceae	March - June	White
Alstonia macrophylla	-	Apocynaceae	July - Sep	white
Alstonia scholaris	Devil's tree	Apocynaceae	Autum	Greenish tree, Scented
Anogeissus acuminate	-	Combrataceae	Feb-March	Yellowish white, fragrant
Anthocephalus chinensis	Kadamba	Rubiaceae	June to August	Orange, scented, small silk
Araucaria bidwillii	Bunya bunya	Araucariaceae	-	-
Araucaria cookii	New Caledonia oine	Araucariaceae	-	-
Araucaria cunninghamii	Moreton Bay pine/ Hoop pine	Araucariaceae	-	-
Berria ammonilla	-	Tiliaceae	May - June	White
Brownea grandiceps	Rose of venezulea	Caesalpiniaceae	March - May	Reddish purple
Cananga odorata	Ylang-ylang	Annoaceae	May - Oct	Pale – yellow fragrant
Cassia moschata	Bronze shower tree	Caesalpiniaceae	March	Yellow turning brick – ed with age, sepals reddish
Casuarina equisetifolia	Beefwood/Australian oak	Casuarinaceae	-	Pinkish
Cedrela toona	Indian mahogany	Meliaceae	Feb - March	White fragrant
Ceiba pentandra	Kapok tree	Bombacaceae	Jan - March	Creamy white (or) light pink

Dalbergia lanceolaria	-	Caesalpiniaceae	-	-
Dillenia indica	Elephant apple	Dilleniaceae	June - July	White, sweet scented
Dubanga sonneratioides	-	Sonneratiacae	-	White
Eucalyptus calophylla	-	Myrtaceae	-	White (or) cream colour
Eucalyptus citriodora	Lemon scented eucalyptus	Myrtaceae	March - May	white
Eucalyptus maculata	-	Myrtaceae	April - May	-
Firmiana colorata	Bonefire tree	Sterculiaceae	Mar - April	Scarlet
Grevillea robusta	Silver Oak	Proteaceae	Mar - April	Orange
Inga dulcis	Manilla tamarind	Euphorbiaceae	Jan - April	-
Lagerstroemia parviflora	-	Lythraceae	Summer	White
Markhamia platycalyx	-	Bignoniaceae	Feb - Mar	Yellow with red lines
Pachira cyathophora	-	Bombacaceae	Mar - May	Cream - white
Parmentiera edulis	Guajilote tree	Bignoniaceae	-	Greenish yellow
Polyalthia longifolia	Mast tree/ Indian fir	Annonaceae	Mar - April	Greenish yellow
Pterygota alata	Buddha's coconut	Sterculiaceae	Feb - March	Greenish purple (or) rusty brown outside & red veined within
Sterculia villosa	Dung tree	Sterculiaceae	-	Yellow with purple centre
Stereospermum kunthianum	-	Bignoniaceae	April - June	Dark purple changing to pale
Swietenia macrophylla	Large leaved mahogany	Meliaceae	-	White
Tabebuia rosea	Rosy trumpet tree	Bignoniaceae	-	Rosy - pink
Taxodium distichum	Bald cypress/ Swamp cypress	Pinaceae	-	Purple

Tectona grandis	Teak	Verbenaceae	July - Aug	White
Terminalia catappa	Indian almond	Combretaceae	Mar - April	White
Tipuana tipu	Pride of Bolivia/Rosewood	Caesalpiniaceae	Several months from summer	Bright golden

38. Trees of medium-sized nature (8 to 15 m)

Botanical Name	**Common name**	**Family**	**Flowering season**	**Flower colour**
Acacia auriculiformis	Golden shower	Mimosaceae	Sep-Oct	Yellow scented
Acacia suma	-	Mimosaceae	Rainy & cold season	white
Amherstia nobilis	Queen of flowering trees	Caesalpiniaceae	Feb-April	Petals brilliant red, edged with yellow
Azadirachta indica	Neem tree	Meliaceae	March-April	White, honey scented
Caesalpinia cacalaco	-	Caesalpiniaceae	-	Petals yellow or orange marked with spots
Cassia fistula	Golden shower	Caesalpiniaceae	Early summer to Sep	Yellow
Cassia javanica	Java cassia	Caesalpiniaceae	April	Bright pink
Cassia renigera	Burmese pink cassia	Caesalpiniaceae	April - May	Pink scented
Cassia siamea	Kassod tree	Caesalpiniaceae	June – Jan Peak in autumn	Yellow
Cochlospermum religiosum	Yellow silk otton	Cochlospermaceae	Feb - March	Brilliant yellow
Erythrina variegata	Indian coral tree	Caesalpiniaceae	Feb - March	Red

Guaiacum officinale	Tree of Life	Zygophyllaceae	-	Blue
Jacaranda mimosifolia	Mimosa – leaved jacaranda	Bignoniaceae	Mar - June	Bluish purple
Klenihovia hospita	Tree antigonon	Sterculiaceae	July - Sep	Rose coloured
Leucaena leucocephala	Horse tamarind	Mimosaceae	Mar – April & again Aug - Nov	White
Melia azedarach	Indian lilac/China berry	Meliaceae	-	Purple in colour, scented
Mesua ferrea	Iron-wood tree	Guttiferae	Feb - May	White, fragrant
Millettia peguensis	Jewels on a string	Caesalpiniaceae	Mar - April	Red (or) purplish
Millettia thonningii	-	Caesalpiniaceae	Mar - April	Reddish violet
Mimusops elengi	Indian Medlar	Sapotaceae	Spring & through-out warm season	Off – white fragrant
Pterospermum semisagittatum	-	Sterculiaceae	-	White
Pterospermum suberifolium	-	Sterculiaceae	Summer	Yellowish white scented
Putranjiva roxburghii	Child-life tree/ Lucky bean tree	Sterculiaceae	Mar - May	Yellow
Salix babylonica	-	Saliaceae	-	-
Saraca asoca	Asoka tree	Caesalpiniaceae	Mar - May	Orange red
Schotia latifolia	Elephant hedge bean tree	Caesalpiniaceae	Summer	Dark crimson
Syzygium malaccense	Malay rose-apple	Myrtaceae	Summer & Nov - Dec	Crimson
Wrightia tomentosa	-	Apocynaceae	April - July	Vermillion

39. Trees with dwarf nature (4 to 8m)

Botanical Name	Common name	Family	Flowering season	Flower colour
Callistemon pollandi	-	Myrtaceae	Feb-April	-
Cerbera tanghin	-	Apocynaceae	-	White
Fugosia hakaefolia	-	Malvaceae	June	Yellow, a red purple spot at the base
Lagerstroemia thorellii	-	Lythraceae	June - Sep	Mauve & fading to almost
Polyalthia coffeoides	-	Annonaceae	-	Greenish yellow

40. Trees suitable for pot culture

Botanical Name	Common name	Family	Flowering season	Flower colour
Araucaria cookii	New Caledonia oine	Araucariaceae	-	-
Araucaria hetrophylla	Norfolk Island pine/ Indoor fir	Araucariaceae	-	-
Brassaia actinophylla	Queensland umbrella tree	Araliaceae	-	Red
Cycas circinalis	Fern palm	Cycadaceae	-	-
Ficus benjamina	Java fig	Moraceae	-	-
Ixora parviflora	Torch tree	Rubiaceae	Mar - May	White, rarely pink, scented
Pinus longifolia	Long-leaved pine	Pinaceae	-	-
Podocarpus macrophylla	Buddhist pine	Podacarpaceae	-	-
Schotia latifolia	Elephant hedge bean tree	Caesalpiniaceae	Summer	Dark crimson

41. Trees with hardy nature

Botanical Name	Common name	Family	Flowering season	Flower colour
Azadirachta indica	Neem tree	Meliaceae	March-April	White, honey scented
Cassia fistula	Golden shower	Caesalpiniaceae	Early summer to Sep	Yellow
Cupressus sempervirens	Italian cypress	Cupressaceae	-	Yellow
Saraca asoca	Asoka tree	Caesalpiniaceae	Mar - May	Orange red
Swietenia macrophylla	Large leaved mahogany	Meliaceae	-	White
Syzygium cumini	Indian blackberry/ Java plum	Myrtaceae	Mar - June	White, fragrant
Tamarindus indica	Tamarind	Caesalpiniaceae	May - Aug	Red & yellow

42. Trees which belongs to the family Annonaceae are

S.No.	Botanical Name	Common name	Flowering season	Flower colour
1.	*Monodora grandiflora*	African orchid	Nov - Dec	Green with brown margin
2.	*Polyalthia coffeoides*	-	-	Greenish yellow
3.	*Polyalthia longifolia*	Mast tree/ Indian fir	Mar - April	Greenish yellow
4.	*Polyalthia longifolia* 'Angustifolia'	-	-	-
5.	*Polyalthia longifolia* 'Pendula'	-	-	-
6.	*Polyalthia suberosa*	-	-	Yellow (or) Reddish brown

43. Trees which belongs to the family Sterculiaceae are

S.No.	Botanical Name	Common name	Flowering season	Flower colour
1.	*Firmiana colorata*	Bonefire tree	Mar - April	Scarlet
2.	*Heritiera macrophylla*	-	Mar - April	Pinkish rose
3.	*Klenihovia hospita*	Tree antigonon	July - Sep	Rose coloured
4.	*Pterospermum acerifolium*	Maple twist/Muchkunda	Spring & early summer	Pure white, scented
5.	*Pterospermum semisagittatum*	-	-	White
6.	*Pterospermum suberifolium*	-	Summer	Yellowish white scented
7.	*Pterygota alata*	Buddha's coconut	Feb - March	Greenish purple (or) rusty brown outside & red veined within
8.	*Pterygota alata* 'Diversifolia'	-	-	-
9.	*Putranjiva roxburghii*	Child-life tree/Lucky bean tree	Mar - May	Yellow
10.	*Sterculia villosa*	Dung tree	-	Yellow with purple centre

44. Trees which belongs to the family Apocynaceae are

S.No.	Botanical Name	Common name	Flowering season	Flower colour
1.	*Alstonia macrophylla*	-	July - Sep	White
2.	*Alstonia scholaris*	Devil's tree	Autum	Greenish tree, Scented
3.	*Cerbera manghas*	-	Throughout the year but profusely during summer	White with pink centre, fragrant
4.	*Cerbera tanghin*	-	-	White
5.	*Holarrhena antidysenterica*	Easter tree/Ivory tree	-	White (or) creamy scented
6.	*Holarrhena floribunda*	-	-	-
7.	*Plumeria alba*	Pagoda tree	-	White (or) various shades of pink, yellow, red (scented)
8.	*Plumeria lutea*	-	-	-
9.	*Plumeria obtusa*	-	-	-
10.	*Plumeria rubra*	-	-	-
11.	*Plumeria rubra* forma bicolor	-	-	-
12.	*Plumeria rubra* forma tricolor	-	-	-
13.	*Plumeria rubra* forma typica	-	-	-
14.	*Thevetia peruviana*	Yellow oleander	Throughout the year	Yellow
15.	*Wrightia tomentosa*	-	April - July	Vermillion

45. Trees which comes under the family Araucariaceae are

S.No.	Botanical Name	Common name	Flowering season	Flower colour
1.	*Araucaria bidwillii*	Bunya bunya	-	-
2.	*Araucaria cookii*	New Caledonia oine	-	-
3.	*Araucaria cunninghamii*	Moreton Bay pine/Hoop pine	-	-
4.	*Araucaria hetrophylla*	Norfolk Island pine/Indoor fir	-	-

46. Trees which belongs to the family Bignoniaceae are

S.No.	Botanical Name	Common name	Flowering season	Flower colour
1.	*Crescentia cujete*	Begger's bowl	Rainy season	Yellowish with purplish veins
2.	*Dolichandrone spathacea*	-	March - May	White
3.	*Heterophragma adenophyllum*	-	June - Aug	Corolla is of yellowish brown tube, swollen at the top
4.	*Heterophragma sp.*	-	-	-
5.	*Jacaranda filicifolia*	Fern leaf Jacaranda	Feb - Mar	Bluish purple
6.	*Jacaranda mimosifolia*	Mimosa – leaved jacaranda	Mar - June	Bluish purple
7.	*Kigelia pinnata*	Sausage tree/cucumber tree	Mar - May	Deep chocolate red, evil smelling
8.	*Markhamia platycalyx*	-	Feb - Mar	Yellow with red lines
9.	*Milligtonia hortensis*	Tree jasmine/Indian cork tree	-	White with greenish tinge, fragrant
10.	*Oroxylum indicum*	Indian trumpet flower	June - July	Lurid - purple
11.	*Parmentiera cereifera*	Candle tree	May - Aug	White (or) dull pink

12.	*Parmentiera edulis*	Guajilote tree	-	Greenish yellow
13.	*Spathodea campanulata*	African tulip tree/Scarlet bell tree	Jan - Feb	Orange crimson outside, having yellow wrinkled margin
14.	*Stereospermum kunthianum*	-	April - June	Dark purple changing to pale purple
15.	*Tabebuia argentea*	Golden bell	Feb - April	Bright yellow
16.	*Tabebuia avellanedae*	-	Winter	Reddish mauve
17.	*Tabebuia chrysantha*	-	Feb - Mar	Bright yellow
18.	*Tabebuia donnell-smithii*	Gold tree	Mar - May	Bright yellow
19.	*Tabebuia pallida*	-	Mar - May	Light pink & yellow outside
20.	*Tabebuia palmeri*	-	Nov - Jan	Lavender rose with white throat, striped yellow & purple
21.	*Tabebuia rosea*	Rosy trumpet tree	-	Rosy - pink

47. Trees which belongs to the family Bombacaceae are

S.No.	Botanical Name	Common name	Flowering season	Flower colour
1.	*Adansonia digitata*	Monkey bread/Sour gourd	June/July	Creamy white, unpleasant smell
2.	*Bombax ceiba*	Red silk cotton tree	February	Red
3.	*Bombax ellipticum*	Shaving brush tree	-	Purple
4.	*Bombax ellipticum* 'Album'	White shaving brush tree	-	-
5.	*Ceiba pentandra*	Kapok tree	Jan - March	Creamy white (or) light pink
6.	*Chorisia speciosa*	Floss silk tree	October	Pinkish outer surface, lower half of inner surface pinikish-white gradually fading
7.	*Pachira cyathophora*	-	Mar - May	Cream - white

48. Trees which belongs to the family Caesalpiniaceae are

S.No.	Botanical Name	Common name	Flowering season	Flower colour
1.	*Amherstia nobilis*	Queen of flowering trees	Feb-April	Petals brilliant red, edged with yellow
2.	*Andira inermis*	Cabbage tree	March-April	Purple
3.	*Baikiaea insignis*	-	April - June	Snow white with lower petal (or) lip lemon yellowFlower buds are large, dark brown& velvety
4.	*Bauhinia blakeana*	Hongkong orchid tree	Oct-Mar	Rich reddish (or) rose-purple almost crimson
5.	*Bauhinia purpurea*	Camel's foot tree	October	-
6.	*Bauhinia variegata*	Purple orchid tree	Jan - March	Purple pink
7.	*Bahuhinia variegate* var. Candida	-	-	-
8.	*Brownea ariza*	-	Feb - June	-
9.	*Brownea coccinea*	West Indian mountain rose	Feb-March & also in September	Pink
10.	*Brownea grandiceps*	Rose of venezulea	March - May	Reddish purple
11.	*Brownea hybrida*	-	Summer months	Reddish purple
12.	*Brownea macrophylla*	-	-	Orange Scarlet
13.	*Butea monosperma*	Flame of forest	February	Flaming orange
14.	*Caesalpinia cacalaco*	-	-	Petals yellow or orange marked with spots

15.	*Caesalpinia coriaria*	Divi Divi	April -July	Greenish yellow (scented)
16.	*Cassia excelsa*	Crown-of-gold tree	Early summer	Bright yellow
17.	*Cassia fistula*	Golden shower	Early summer to Sep	Yellow
18.	*Cassia fistula* (Dwarf type)	-	-	-
19.	*Cassia grandis*	Horse cassia/Pink shower	Feb - April	Pale pink
20.	*Cassia hybrida*	-	-	-
21.	*Cassia lancasteri*	-	-	-
22.	*Cassia javanica*	Java cassia	April	Bright pink
23.	*Cassia marginata*	Red cassia	June -July	Terracotta
24.	*Cassia moschata*	Bronze shower tree	March	Yellow turning brick – ed with age, sepals reddish
25.	*Cassia multijuga*	-	Sep - Nov	Yellow
26.	*Cassia nodosa*	Pink cassia	March - April	Bright pink
27.	*Cassia renigera*	Burmese pink cassia	April - May	Pink scented
28.	*Cassia siamea*	Kassod tree	June – Jan Peak in autumn	Yellow
29.	*Cassia spectabilis*	-	Sep - Dec	Golden yellow
30.	*Castanospermum australe*	Moreton Bay chestnut	March	Yellow to orange
31.	*Clitoria arborea*	-	May - June	Shades of violet, centre deeper with red markings
32.	*Colvillea racemosa*	Colville's Glory	Sep - Oct	Orange
33.	*Dalbergia lanceolaria*	-	-	-
34.	*Dalbergia sissoo*	South Indian Rewood	-	Yellowish white scented

35.	*Delonix elata*	-	Aug - March	White but turning yellow before fading
36.	*Delonix regia*	Flame of forest	April - Aug	Orange -scarlet
37.	*Erythrina variegata*	Indian coral tree	Feb - March	Red
38.	*Gliricidia sepium*	Mata-raton	-	Pink, white, purpe
39.	*Millettia peguensis*	Jewels on a string	Mar - April	Red (or) purplish
40.	*Millettia thonningii*	-	Mar - April	Reddish violet
41.	*Parkinsonia aculeata*	Jerusalem thorn	Summer months	Yellow
42.	*Peltophorum africanum*	-	Light yellow	June - Aug
43.	*Peltophorum brasiliense*	-	-	Yellow
44.	*Peltophorum pterocarpum*	Rusty shield bearer	Mar – June & also in Sep - Nov	Yellow
45.	*Piscidia erythrina*	Jamaica dog-wood	April	Summer
46.	*Pongamia pinnata*	Indian beech/Poonga oil plant	-	Rose (or) liliac
47.	*Pterocarpus indicus*	Burmese rose-wood/Padauk	-	Golden yellow fragrant
48.	*Saraca asoca*	Asoka tree	Mar - May	Orange red
49.	*Saraca cauliflora*	-	Mar - May	Yellow & May
50.	*Saraca declinata*	Red saraca	Late Feb - April	-
51.	*Saraca thaipingensis*	Yellow saraca	Jan - Mar	Golden yellow
52.	*Schotia latifolia*	Elephant hedge bean tree	Summer	Dark crimson
53.	*Sesbania grandiflora*	Swamp pea	Sep - Nov	White, red, pink
54.	*Tipuana tipu*	Pride of Bolivia/Rosewood	Several months from summer	Bright golden

49. Trees which belongs to the family Combretaceae are

S.No.	Botanical Name	Common name	Flowering season	Flower colour
1.	*Anogeissus acuminate*	-	Feb-March	Yellowish white, fragrant
2.	*Terminalia arjuna*	White murdah	April - May	Dull yellow
3.	*Terminalia catappa*	Indian almond	Mar - April	White

50. Trees which comes under the family Cupressaceae are

S.No.	Botanical Name	Common name	Flowering season	Flower colour
1.	*Cupressus sempervirens*	Italian cypress	-	Yellow
2.	*Juniperus chinensis*	Chinese juniper	-	-
3.	*Thuja orientalis*	Chine thuja	-	Bluish green

51. Trees which belongs to the family Malvaceae are

S.No.	Botanical Name	Common name	Flowering season	Flower colour
1.	*Fugosia hakaefolia*	-	June	Yellow, a red purple spot at the base
2.	*Hibiscus tiliaceus*	Cork wood/Yellow mallow tree	Summer & rainy	Yellow with crimson eye
3.	*Thespesia populnea*	Indian tulip tree/Portia tree	April - May	Yellow & turn purplish with age

52. Trees which belongs to the family Mimosaceae are

S.No.	Botanical Name	Common name	Flowering season	Flower colour
1.	*Acacia arabica*	Indian gum-arabic tree	Rainy season	Yellow scented
2.	*Acacia auriculiformis*	Golden shower	Sep-Oct	Yellow scented
3.	*Acacia baileyana*	-	Winter end (or) Early spring	Yellow
4.	*Acacia decurrens*	Queen wattle/Green wattle	Winter end (or) Early spring	Yellow
5.	*Acacia suma*	-	Rainy & cold season	white
6.	*Acacia sp.*	-	-	-
7.	*Adenanthera pavonina*	Red sandalwood/Bead tree	End of March – April Sometimes in Sep-Oct	Yellow & Scented
8.	*Albizzia chinensis*	-	-	white
9.	*Albizzia julibrission*	Pink siris/Silk tree	-	-
10.	*Albizzia lebbek*	Black siris/Parrot tree	-	Creamy white, fragrant
11.	*Albizia procera*	White siris	May - Sep	Greenish white
12.	*Albizzia richardiana*	-	-	White
13.	*Aleurietes moluccana*	Candle nut/Varnish tree	March - June	White
14.	*Haematoxylon campechianum*	Logwood/Campeachy tree	-	Yellow scented
15.	*Leucaena leucocephala*	Horse tamarind	Mar – April & again Aug - Nov	White
16.	*Parkia biglandulosa*	Ball batmitton tree	Nov - Dec	Brown & velvety but becomes white as the flowers open

17.	*Prosopis julifora*	Algaoba bean/Mesquite	April - May	Greenish turning yellow, fragrant
18.	*Samanea saman*	Rain tree	April - Oct	Yellow

53. Trees which belongs to the family Moraceae are

S.No.	Botanical Name	Common name	Flowering season	Flower colour
1.	*Artocarpus altilis*	Bread fruis	-	-
2.	*Ficus bengalensis*	Banyan tree	-	-
3.	*Ficus bengalensis* 'Krishnae'	-	-	-
4.	*Ficus benjamina*	Java fig	-	-
5.	*Ficus elastica*	Indian Rubber tree	-	-
6.	*Ficus hispida*	-	-	-
7.	*Ficus infectoria*	-	-	-
8.	*Ficus religiosa*	Pipal tree	-	-
9.	*Ficus rumphii*	-	-	-

54. Trees which belongs to the family Rubiaceae are

S.No.	Botanical Name	Common name	Flowering season	Flower colour
1.	*Anthocephalus chinensis*	Kadamba	June to August	Orange, scented, small silk
2.	*Gardenia latifolia*	Cape jasmine	Summer	White, highly scented
3.	*Ixora parviflora*	Torch tree	Mar - May	White, rarely pink, scented
4.	*Nauclea orientalis*	-	Summer months	Fragrant

55. Trees which belongs to the family Sapindaceae are

S.No.	Botanical Name	Common name	Flowering season	Flower colour
1.	*Filicium decipiens*	Fern-leaf tree	-	white
2.	*Schleichera oleosa*	Lac tree/ Ceylon oak	Spring	Yellow

56. Trees which belongs to the family Lecythidaceae are

S.No.	Botanical Name	Common name	Flowering season	Flower colour
1.	*Barringtonia acutangula*	Indian Oak	April - May	Pink to Red
2.	*Barringtonia asiatica*	Fish poison tree	July to Sep	White
3.	*Barringtonia racemosa*	Bottle brush oak	Summer & rainy season	Cream coloured
4.	*Careya arborea*	Wild guava	Early summer	Pink & white (foul smell)
5.	*Couroupita guianensis*	Cannon ball tree	Several times in a year	Outside a mixture of white, yellow & inside deep pink (or) crimson
6.	*Gustavia augusta*	Stink wood	Mar - Sep	White (or) tinged pink fragrant
7.	*Napoleona imperialis*	-	Feb - June	Brown, rose & yellowish

57. Trees which belongs to the family Tiliaceae are

S.No.	Botanical Name	Common name	Flowering season	Flower colour
1.	*Berria ammonilla*	-	May - June	White
2.	*Muntingia calabura*	Chinese cherry	-	White

58. Trees which comes under the family Euphorbiaceae are

S.No.	Botanical Name	Common name	Flowering season	Flower colour
1.	*Bishofia javanica*	Toog tree/ Java cedar	Feb-April	Pale green
2.	*Hura crepitans*	Sandbox tree	-	-
3.	*Inga dulcis*	Manilla tamarind	Jan - April	-
4.	*Sapium sebiferum*	Chinese tallow tree	April – June & Oct - Nov	-

59. Trees which belongs to the family Myrtaceae are

S.No.	Botanical Name	Common name	Flowering season	Flower colour
1.	*Callistemon citrinus*	Red bottle-brush	-	Calyx pale greenish, filaments numerous, free, bright red
2.	*Callistemon pollandii*	-	Feb-April	-
3.	*Eucalyptus calophylla*	-	-	White (or) cream colour
4.	*Eucalyptus citriodora*	Lemon scented eucalyptus	March - May	white
5.	*Eucalyptus globulus*	Tasmanian blue gum	Dec - May	-
6.	*Eucalyptus maculata*	-	April - May	-
7.	*Melaleuca leucadendran*	White bottle brush/ Swamp tea tree	Hot & rainy seasons	whitish
8.	*Syzygium cumini*	Indian blackberry/ Java plum	Mar - June	White, fragrant
9.	*Syzygium malaccense*	Malay rose-apple	Summer & Nov - Dec	Crimson

60. Trees which belongs to the family Meliaceae are

S.No.	Botanical Name	Common name	Flowering season	Flower colour
1.	*Azadirachta indica*	Neem tree	March-April	White, honey scented
2.	*Cedrela microcarpa*	-	-	white
3.	*Cedrela toona*	Indian mahogany	Feb - March	White fragrant
4.	*Melia azedarach*	Indian lilac/China berry	-	Purple in colour, scented
5.	*Swietenia macrophylla*	Large leaved mahogany	-	White
6.	*Swietenia mahagoni*	Spanish mahogany	Summer	Green

61. Trees which belongs to the family Sapotaceae are

S.No.	Botanical Name	Common name	Flowering season	Flower colour
1.	*Chrysophyllum cainito*	Star apple	-	Purplish white
2.	*Mimusops elengi*	Indian Medlar/ Mahilampoo	Spring & throughout warm season	Off – white fragrant

62. Trees which belongs to the family Verbenaceae are

S.No.	Botanical Name	Common name	Flowering season	Flower colour
1.	*Citharexylum quadrangulare*	Zither-wood	March - May	White scented
2.	*Tectona grandis*	Teak	July - Aug	White

63. Ornamental trees with **round or globular canopy** are *Ficus infectoria, Mimusops elengi, Dillenia indica, Plumeria alba, Morus nigra*

64. Ornamental trees with **drooping/ pendulous growth habit** are *Anthocephalus cadamba, Salix babylonica, Polyalthia longifolia, Cupressus forbessi, Callistemon lanceolatus*

65. **White coloured flowering trees** are *Alstonia scholaris, Azadirachta indica, Callistemon salignuso, Cedrela toona, Millingtonia hortensis, Mimusops elengi, Tectonia grandis, Terminalia catappa, Dillenia indica*

66. Tree species which produce **blue flowers** are *Jacaranda acutifolia* & *Guaicum officinale*

67. **Yellow coloured** flowering trees are *Callistemon polandii, Cassia fistula, Cassia siamea, Heterophragma adenophyllum, Michelia champaka, Parkinsonia aculeata, Pterospermum acerifolium, Putranjiva roxburghii, Terminalia arjuna, Thespesia populnea*

68. **Red coloured** flowering trees are *Bassia latifolia, Callistemon lanceolatus, Spathodea campanulata*

69. Tree species which produce **scarlet red** colour flowers are *Saraca asoka, Brownea grandiceps*

70. Tree species which produce **purple colour** flowers *Bauhinia purpurea, Bauhinia blackeana, Milletia ovalifolia, Lagerstroemia speciosa, Melia azedarach*

71. **Pink colour** flowering trees are *Cassia nodosa, Chorisia speciosa, Cassia javanica, Cassia renigera, Pongamia pinnata*

72. Ornamental trees with **beautiful trunk** are *Eucalyptus hybrida, Adansonia digitata, Bombax ceiba, Chorisia speciosa, Platanus orientalis*

73. Ornamental trees with **fine foliage** texture are *Grevillea robusta, Cupressus arizonica, Podocarpus heterophyllus, Cryptomeria japonica, Acacia mollissima*

74. **Ornamental foliage trees** which are suitable for **hilly areas** are *Pinus wallichiana, Podocarpus chinensis, Thuja orientalis, Platanus orientalis, Cupressus sempervirens, Juniperus chinensis*

75. **Ornamental foliage trees** which are suitable for **tropical areas** are *Averrahoa carambola, Callistemon lanceolatus, Citharexylum subserratum, Eucalyptus* spp., *Phyllanthus emblica, Putranjiva roxburghii, Terminalia arjuna, Polyalthia longifolia, CLusea rosea and Filicium decipens*

76. **Ornamental trees** suitable for planting in **moisture rich areas** are *Putranjiva roxburghii, Alstonia scholaris, Cassia javanica, Callistemon lanceolatus, Lagerstroemia speciosa*

77. Ornamental trees suitable for **home garden** are *Tecomella undulata, Plumeria rubra, Brownea coccinea, Cassia javanica, Acacia auriculiformis, Cassia fistula*

78. Ornamental trees used for planting as **shade tree** are *Kigelia pinnata, Pongamia pinnata, Azadirachta indica, Ficus benghalensis, Michelia champaca, Diospyros embryopteris, Ficus infectoria, Ficus retusa* and *Tamarindus indica*

79. Ornamental trees which **check air pollution** are *Morus* spp, *Poplar hybrida, Plumeriaacutifolia, Ficus infectoria*

80. Ornamental trees used for **screening purpose** are *Grevillea robusta, Eucalyptus* spp, *Poplar* spp, *Polyalthia longifolia, Polyalthia pendula, Cupressus sempervirens, Casuarina equisetifolia, Putranjiva roxburghii*

81. Ornamental trees suitable for **alkaline & saline soils** are *Casuarina equisetifolia, Cassia fistula, Parkinsonia aculeata*

82. Ornamental trees which possess **thorns** are *Erythrina indica, Bombax ceiba, Chorisia speciosa, Parkinsonia aculeata, Acacia nilotica*

83. Ornamental trees which produce **fragrant flowers** are *Magnolia grandiflora, Moringa oleifolia, Plumeria alba, Alstonia scholaris, Gardenia latifolia, Dillenia indica, Mimusops elengi, Michalia champaca* and *Pterospermum acerifolium*

84. Ornamental trees suitable for planting as **wind break** are *Delonix regia, Tabebuia speciosa, Peltophorum ferrugineum, Pterospermum acerifolium, Eugenia jambolana*

85. **Ever- blooming ornamental trees** are *Mimusops elengi, Callistemon lanceolatus, Solanum macranthum, Thespesia populnea, Plumeria acuminata*

86. Ornamental trees valued for **fruits** are *Koelreuteria paniculata, Heterophragma adenophyllum, Kigelia pinnata*

87. Ornamental trees suitable for planting in **dry areas** are *Cassia fistula, Erythrina indica, Acacia nilotica, Acacia mollissima, Butea monosperma, Albizia lebbeck, Casuarina equisetifolia, Melia azedarach, Prosopsis juliflora*

88. Ornamental trees of **Indian origin** are Asoka, Pipal, Banyan, Nagkesar (*Mesua ferrea*), Palas (*Butea monosperma*), Kadam, *Michelia champaca*, Neem, Arjun tree, Kachnar (*Bauhinia variegata*), Karanj (*Pongamia glabra*), *Thespesia populnea, Dillenia indica* (Chalta), *Cratavea*

roxburghi (Barna), *Sterculia colorata*, *Bassia latifolia* (Mahua), *Cassia fistula* (Amaltas), *Cedrella toona* (Indian mahogany tree), *Erythrina variegata* (Pangara), *Heterophragma heterophyllum* (Monkey tail), *Lagerstroemia speciosa* (Pride of India), *Terminalia catappa* (False Almond)

89. **Flowering time of ornamental trees**

March:*Bauhinia alba*, *Bauhinia variegata*, Semal (*Bombax ceiba*), Flame of forest (*Butea monosperma*)

March- April: Tree of life (*Guaiacum officinale*), Teak (*Tectona grandis*), Jiva putra(*Putranjiva roxburghii*)

April: Indian Mahogani (*Swietenia mahagoni*), Pangara (*Erythrina indica*), Silver oak (*Grevillea robusta*), Neel Gulmohar (*Jacaranda mimosaefolia*), Rose wood (*Dalbergia sissoo*), Pagoda tree (*Plumeria* spp.), Sita Ashoka (*Saraca asoca*)

April- May: Neem (*Azadirachta indica*), Siris (*Albizia lebbeck*), Akash neem (*Millingtonia hortensis*), Jerusalem thorn (*Parkinsonia aculeata*), Kanak champa (*Pterospermum acerifolium*), Tree of Heaven (*Ailanthus excelsa*)

April- August: Queen of flowering trees (*Lagerstroemia speciosa*)

May: Swarna champa (*Michelia champaca*), Karanj (*Pongamia pinnata*), Badam (*Terminalia catappa*)

May- June: Gulmohar (*Delonix regia*), Balam khira (*Kigelia Pinnata*)

May-July: Amaltas (*Cassia fistula*), Arjun tree (*Terminalia arjuna*)

May- August: Elephant apple (*Dillenia indica*)

July- August: Devil's tree, Monkey bread tree (*Adansonia digitata*)

July- January: Kassod tree (*Cassia siamea*)

90. Kadam tree is associated with **Lord Krishna**

91. Semal tree is associated with **Lord Shiva**

92. Amaranthus tree is associated with **Lord Kali**

93. Yellow Amaltas tree is associated with **prosperity in trade**

94. Ashoka, Sal & Palash tree are associated with **Buddha**

II. Fill in the blanks

1. Semal tree is associated with..............
2. Amaranthus tree is associated with..............
3. Yellow Amaltas tree is associated with..............
4. *Ixora parviflora* is commonly called as..............
5. *Callistemon lanceolatus* produce ....coloured flowers
6. Gulmohar produce flower during the month of..............
7. *Guaicum officinale* produce flower in ....colour
8. Bauhinia is commercially propagated by..............
9. *Putranjiva roxburghii* produce flower in ....colour
10. *Millingtonia hortensis* produce flower in ....colour

III. Choose the best answer

1. Tree which is suitable for marshy area is ....

   Hibiscus tiliaceus *Mimusops elengi* *Polyalthia longifolia*
2. Tree species which produce scented flower is ....

   Hibiscus tiliaceus *Mimusops elengi* *Polyalthia longifolia*
3. Tree which is used as a specimen tree is ....

   Prosopsis juliflora *Polyalthia longifolia* *Grevillea robusta*
4. Ornamental tree which is used for screening purpose is..............

   Putranjiva roxburghii *Hibiscus tiliaceus* *Mimusops elengi*
5. Ornamental tree which grows well in alkaline & saline soil is ....

   Polyalthia longifolia *Casuarina equisetifolia* *Putranjiva roxburghii*
6. Ornamental tree which is of Indian origin is..............

   Terminalia catappa *Prosopsis juliflora* *Polyalthia longifolia*
7. Ornamental tree which produce flowers during the month of May is ....

   Devil's tree Kassod tree Swarna champa

8. Tree which is not grazed by cattle is..............

   Cassia fistula *Terminalia catappa* *Prosopsis juliflora*

9. Tree which is suitable for arid & rocky region is..............

   Albizia lebbek *Heterophragma adenophyllum* *Erythrina variegata*

10. Ornamental tree which is suitable for planting in dry area is ....

    Melia azedarach *Barringtonia racemosa* *Salix tetrasperma*

11. Ever- blooming ornamental tree is..............

    Thespesia populnea *Heterophragma adenophyllum* *Kigellia pinnata*

12. Ornamental tree which is suitable for planting as wind break is ....

    Peltophorum ferrugineum *Magnolia grandiflora* *Moringa oleifolia*

13. Ornamental tree which produce fragrant flower is ....

    Peltophorum ferrugineum *Alstonia scholaris* *Pterospermum acerifolium*

14. Ornamental tree which possess thorns is..............

    Parkinsonia aculeata *Peltophorum ferrugineum* *Pterospermum acerifolium*

15. Tree which is suitable for avenue planting is..............

    Madhuca longifolia *Millingtonia hortensis* *Mimusops elengi*

IV. Match the following

1.	*Cassia javanica*	:	Moist areas
2.	*Ficus tomentosa*	:	*Monodora grandiflora*
3.	*Salix babylonica*	:	Bonfire tree
4.	*Barringtonia acutangula*	:	Not grazed by cattle
5.	African orchid	:	Rocky regions
6.	*Firmiana colorata*	:	Marshy areas
7.	*Caesalpinia coriaria*	:	Peacock flower
8.	*Samanea saman*	:	Divi-divi
9.	*Thevetia peruviana*	:	Rain tree
10.	*Delonix regia*	:	Yellow oleander

Answers

II. Fill in the blanks

1. Shiva
2. Kali
3. Prosperity in trade
4. Torch tree
5. Red
6. May- June
7. Blue
8. Seeds
9. Yellow
10. White

III. Choose the best answer

1. *Hibiscus tiliaceus*
2. *Mimusops elengi*
3. *Polyalthia longifolia*
4. *Putranjiva roxburghii*
5. *Casuarina equisetifolia*
6. *Terminalia catappa*
7. Swarna champa
8. *Cassia fistula*
9. *Albizia lebbek*
10. *Melia azedarach*
11. *Thespesia populnea*
12. *Peltophorum ferrugineum*
13. *Alstonia scholaris*
14. *Parkinsonia aculeata*
15. *Madhuca longifolia*

IV. Match the following

1. Not grazed by cattle
2. Rocky regions
3. Marshy areas
4. Moist areas
5. *Monodora grandiflora*
6. Bonfire tree
7. Divi-divi
8. Rain tree
9. Yellow oleander
10. Peacock flower

Annexure I- Trees

S. No.	Botanical Name	Common name	Family	Flowering season	Flower colour	Remarks
1.	*Acacia arabica*	Indian gum-arabic tree	Mimosaceae	Rainy season	Yellow scented	Used as avenue tree
2.	*Acacia auriculiformis*	Golden shower	Mimosaceae	Sep-Oct	Yellow scented	Medium sized, thornless, quick growing, evergreen of 13-15 cm. Propagated by seeds. Used in large scale for tree planting programme during to its capacity for rapid biomass production
3.	*Acacia baileyana*	-	Mimosaceae	Winter end (or) Early spring	Yellow	Well shaped tree growing upto 19m height
4.	*Acacia decurrens*	Queen wattle/ Green wattle	Mimosaceae	Winter end (or) Early spring	Yellow	Propagated by seeds, cuttings, approach grafting
5.	*Acacia suma*	-	Mimosaceae	Rainy & cold season	White	A medium sized tree with
6.	*Acacia sp.*	-	Mimosaceae	-	-	-
7.	*Adansonia digitata*	Monkey bread/ Sour gourd	Bombacaceae	June/July	Creamy white, unpleasant smell	Propagated by seeds. One of the largest tree of the world. Lines for thousand years. Deciduous nature Trunk looks like a bottle shape. Good specimen tree for large

						parts and gardens. Can be planted on side of wide roads
8.	*Adenanthera pavonina*	Red sandal-wood/Bead tree	Mimosaceae	End of March – AprilSometimes in Sep-Oct	Yellow & Scented	Fairly quick growing & well shaped tree. Propagated by large cuttings
9.	*Ailanthus excelsa*	Tree of heaven/ Paradise tree	Simarubaceae	End of cold season	White (or) yellow	Tall, deciduous – quick growing tree. Propagated by seeds, cuttings and suckers
10.	*Albizia chinensis*	-	Mimosaceae	-	white	Quick growing tall tree. Planted on road side. Propagated by seeds
11.	*Albizia julibrission*	Pink siris/ Silk tree	Mimosaceae	-	-	-
12.	*Albizia lebbek*	Black siris/ Parrot tree	Mimosaceae	-	Creamy white, fragrant	Large deciduous tree with spreading branches. Shallow root system. Susceptible to strong winds. Not recommended for planting in along road sides. Attached heavily by bark eating caterpillars. Propagated by seeds.
13.	*Albizia procera*	White siris	Mimosaceae	May - Sep	Greenish white	Excellent roadside plant with tall shapely & beautiful crown. Suitable for group planting near lakes & river

						bands. Propagated by seeds
14.	*Albizia richardiana*	-	Mimosaceae	-	White	Quick growing, spreading tree
15.	*Aleurietes moluccana A. fordii*	Candle nut/ Varnish tree	Mimosaceae	March - June	White	Propagated by seeds & cuttings. Tall evergreen tree. Suitable for growing along roadsides, parks and other public gardens. Beautiful tree with uniform spreading crown
16.	*Alstonia macrophylla*	-	Apocynaceae	July - Sep	White	A tall tree with smooth grey bark and shoot branches.
17.	*Alstonia scholaris*	Devil's tree	Apocynaceae	Autum	Greenish tree, Scented	A tall ever green tree of 15-20m height. Branches spreading in tiers or whorls. It is used as an avenue tree in parks and gardens. Specimen tree inside lawns
18.	*Amherstia nobilis*	Queen of flowering trees	Caesalpin-iaceae	Feb-April	Petals brilliant red, edged with yellow	A medium sized, evergreen tree attaining a height of about 8-12m. Propagation is done by air- layering
19.	*Andira inermis*	Cabbage tree	Caesalpiniaceae		March-April	Purple. A beautiful ever-green tree. Suitable for road-side plantation & in large garden. Propagated by seeds

						or cuttings
20.	*Anogeissus acuminata*	-	Combretaceae	Feb-March	Yellowish white, fragrant	A tall evergreen treeIt possesses crooked trunk, numerous spreading & drooping branches, reminiscent of a weeping willow
21.	*Anthocephalus chinensis*	Kadamba	Rubiaceae	June to August	Orange, scented, small silk	A tall deciduous tree (15 – 20m). In India, this tree has a mythological association with Lord Krishna. It provides dense shade. It is propagated by seed
22.	*Araucaria bidwillii*	Bunya bunya	Araucariaceae	-	-	Tall ever green tree
23.	*Araucaria cookii*	New Caledonia oine	Araucariaceae	-	-	Tall ever green tree. Suitable for pot cultureIt has a peculiar tendency to lean slightly side ways. It produces cone. Each scale of a cone conceals a seed
24.	*Araucaria cunninghamii*	Moreton Bay pine/Hoop pine	Araucariaceae	-	-	Tall, ever green tree. Leaves needle shaped, crowded
25.	*Araucaria hetrophylla*	Norfolk Island pine/Indoor fir	Araucariaceae	-	-	It is a pyramidal tree. Propagated from seeds.

No.	Botanical name	Common name	Family	Flowering season	Flower colour	Remarks
						It is of slow growing nature. Suitable for pot culture & indoor decoration. It is grown in garden, particularly in the lawn as a specimen tree
26.	*Artocarpus altilis*	Bread fruit	Moraceae	-	-	Tree grows upto 10 m height with sticky latexIt is of deciduous natureIt produce big fruits and are of seedless and covered with prickles. Propagated by cuttings and also by seeds. Suitable for planting as shade tree in parks and gardens.
27.	*Azadirachta indica*	Neem tree	Meliaceae	March-April	White, honey scented	A medium-large graceful deciduous tree attaining a height of 10 – 16m. Propagated by seeds. Hardy tree with stand salinity & drought. It is suitable for planting as a shade tree along the road side & also in parks and lawn
28.	*Baikiaea insignis*	-	Caesalpiniaceae	April - June	Snow white with lower petal (or) lip lemon	Evergreen ornamental flowering tree, reaching a height of 10 – 12 m. It is

					yellowFlower buds are large, dark brown& velvety	propagated by seedsIt cannot withstand scorching direct sun
29.	*Barringtonia acutangula*	Indian Oak	Lecythidaceae	April - May	Pink to Red	It is a slow growing tree It can withstand water logging. Propagated by seeds. Suitable for planting in small & medium avenues
30.	*Barringtonia asiatica*	Fish poison tree	Lecythidaceae	July to Sep	White	Evergreen tree attaining a height of 12-15m. Suitable for planting in groups in parks
31.	*Barringtonia racemosa*	Bottle brush oak	Lecythidaceae	Summer & rainy season	Cream coloured	Propagated by seeds A small tree growing upto 8m Branches have numerous large leaf scars. Flowers when fall on the water, remain floating for several hoursIdeal tree for planting in groups near lakes & tanks
32.	*Bauhinia blakeana*	Hongkong orchid tree	Caesalpin-iaceae	Oct-Mar	Rich reddish (or) rose-purple almost crimson	It grows to a height of 8m. Propagated by seeds & cuttings flowers profusely seen in summer location
33.	*Bauhinia purpurea*	Camel's foot tree	Caesalpin-iaceae	October	-	Evergreen tree grows to a height of 10-16m. Propa-

						gated by seeds. Tree is almost bare for short timeIt is planted along the medium avenues, parks, gardens & around the temple
34.	*Bauhinia variegata*	Purple orchid tree	Caesalpin-iaceae	Jan - March	Purple pink	A small deciduous tree reaching a height of 6-8m. Flowers looks like orchid. Propagated from seeds. Suitable for road side planting in gardens, parks, railway plot form, office compound group planting. Can be for specimen tree in lawn
35.	*Bahuhinia variegata* var. Candida	-	Caesalpin-iaceae	-	-	-
36.	*Berria ammonilla*	-	Tiliaceae	May - June	White	A tall, well shaped tree. Suitable for avenue planting. Propagated from seeds.
37.	*Bischofia javanica*	Toog tree/ Java cedar	Euphorbi-aceae	Feb-April	Pale green	A large, round – headed deciduous tree. Flowers are dioecious. Very attractive tree when in flower.
38.	*Bixa orellana*	Annatto tree	Bixaceae	Sep - Dec	Rose, Purple (or) white	A small (4-5m) evergreen tree. Propagated from seedIt can be planted conveniently

						at the background of shrubberies, along small roads in parks & at the corner of small gardens.
39.	*Bombax ceiba*	Red silk cotton tree	Bombacaceae	February	Red	A large deciduous tree attaining a height of 25-30m. Propagated by seed and cuttingsIt can be planted in parks or large gardens. Suitable for planting beside town roads
40.	*Bombax ellipticum*	Shaving brush tree	Bombacaceae	-	Purple	Soft wooded deciduous tree with a greenish trunk
41.	*Bombax ellipticum* 'Album'	White shaving brush tree	Bombacaceae	-	-	-
42.	*Brassaia actinophylla Syn: Schefflera actinophylla*	Queensland umbrella tree	Araliaceae	-	Red	An erect growing evergreen tree growing to a height of 8-10m. Propagated by any size of cutting upto 2m long. It can stand heavy pruning. It is a popular pot plant. Suitable for keeping indoors Grows well in warm, humid climate in both semi-shade & full sun
43.	*Brownea ariza*	-	Caesalpin-iaceae	Feb - June	-	Grown as a specimen tree in large gardens. As the flowers

						are mostly produced inside the canopy, they are visible when seen from below the tree.
44.	*Brownea coccinea*	West Indian mountain rose	Caesalpin-iaceae	Feb-March & also in September	Pink	A small evergreen tree of compact growth, 5 – 7m high. Commonly propagated by layering, grafting. The plant should be grown in places sheltered from hot winds & in partial shadeIn large gardens, they may be planted in groups, parti-cularly inside green lawns
45.	*Brownea grandiceps*	Rose of venezulea	Caesalpin-iaceae	March - May	Reddish purple	A handsome tall evergreen tree with spreading crown, about 10 m high Propagated from seeds & air-layering During early stage of growth, it needs to be protected from dry wind & hot sunIt can be planted in large gardens & parks
46.	*Brownea hybrida*	-	Caesalpin-iaceae	Summer months	Reddish purple	A small evergreen tree to 8 m high with spreading branches Suitable for planting along roads. It

						needs protection from scorching sun & dry heat, especially in the first few years after planting. Propagated from seeds & air layering
47.	*Brownea macrophylla*	-	Caesalpiniaceae	-	Orange Scarlet	It is a small tree grows upto 6m. Propagated by airlayering. It prefers partial light & warm humid atmosphere
48.	*Butea monosperma*	Flame of forest	Caesalpiniaceae	February	Flaming orange	When in flower , the tree looks as if covered with flame. Tree is frost-hardy, resistant to drought, tolerant to salinity. Propagated from seeds. It can be used for planting along medium sized roads & also in parks & gardens
49.	*Caesalpinia cacalaco*	-	Caesalpiniaceae	-	Petals yellow or orange marked with spots	A medium sized tree with stout trunk & spreading branches it prefer warm& humid regions. It is a good tree for medium gardens. It suitable for planting along roadside. Propagated by seed & cutting

50.	*Caesalpinia coriaria*	Divi Divi	Caesalpin-iaceae	April -July	Greenish yellow (scented)	A small deciduous umbrella shaped spreading tree with drooping branchesIt can be used for planting on small. This is used as a shade tree in small gardens
51.	*Callistemon citrinus*	Red bottle-brush	Myrtaceae	-	Calyx pale greenish, filaments numerous, free, bright red	A low evergreen tree, reaching a height of 7-10 m with drooping branchesIt is grown both in moist and dry regions. Propagated from seed, cutting and air-layering.
52.	*Callistemon pollandii*	-	Myrtaceae	Feb-April	-	A dwarf tree or tall shrub, 4-5m high A handsome dwarf tree with pendulous branc-hes. Propagated from seeds, cuttings and air- layering
53.	*Calophyllum inophyllum*	Dilo oil tree	Guttiferae	-	Sepals and petals are in white colour, stamens yellow	A large evergreen tree grows to a height of 12m. It grows well under moist tropical climate. Propagated from seeds (or) cuttings. They are suitable for planting in parks and gardens

54.	*Cananga odorata*	Ylang-ylang	Annoaceae	May - Oct	Pale – yellow fragrant	An evergreen, quick growing tree grows to a height of 15-20m tall, it prefers humid climateIt is propagated easily by seed , it is not recommended for along the road side
55.	*Careya arborea*	Wild guava	Lecythidaceae	Early summer	Pink & white (foul smell)	A large, stout, deciduous treeIt prefers a moist soil and humid atmosphereThe tree is planted in parks
56.	*Cassia excelsa*	Crown-of-gold tree	Caesalpin-iaceae	Early summer	Bright yellow	A handsome tree grows upto 10m with softly pubescent branches
57.	*Cassia fistula*	Golden shower	Caesalpin-iaceae	Early summer to Sep	Yellow	A small to medium sized, deciduous hardy treeIt is drought tolerant. Excellent flowering tree for arid regionsIt is suitable along small & medium roads & also planting in parts and gardens
58.	*Cassia fistula* (Dwarf type)		-	Caesalpiniaceae	-	- -
59.	*Cassia grandis*	Horse cassia/ Pink shower	Caesalpin-iaceae	Feb - April	Pale pink	This tree is larger than most pink cassia
60.	*Cassia hybrida*	-	Caesalpin-iaceae	-	-	-

61.	*Cassia lancasteri*	-	Caesalpin-iaceae	-	-	-
62.	*Cassia javanica*	Java cassia	Caesalpin-iaceae	April	Bright pink	A beautiful medium sized treeIt prefers a mild climate & need to open dry wind. Propagated easily by seeds. Very suitable for town roadsIn parks & gardens it can be used for group plants
63.	*Cassia marginata*	Red cassia	Caesalpin-iaceae	June -July	Terracotta	Small, slow growing, spreading tree with short trunk & drooping branches. Propagated by seeds. This small tree can be used in group planting in large gardens (or) as a specimen plant in the corner of small garden to provide umbrella like shade
64.	*Cassia moschata*	Bronze shower tree	Caesalpin-iaceae	March	Yellow turning brick – ed with age, sepals reddish	Tall tree growing to a height of about 20m. It is a beautiful tree when in flower in leafless condition
65.	*Cassia multijuga*	-	Caesalpin-iaceae	Sep - Nov	Yellow	A small quick growing, rather short lived slender tree (or) a large.

66.	*Cassia nodosa*	Pink cassia	Caesalpiniaceae	March - April	Bright pink	An evergreen tree of 12-16m in height with knotted trunk. The tree looks pretty during flowering season. It is a favourite tree in gardens & parks, as well as along the city roads. It can be used as a specimen tree with its graceful leaves & umbrella like crown. Propagated by seeds.
67.	*Cassia renigera*	Burmese pink cassia	Caesalpiniaceae	April - May	Pink scented	A medium sized deciduous tree reaching a height of about 8-10m with drooping branches. It is a good tree for small gardens it can be used for planting along small roads in parks & townships.
68.	*Cassia siamea*	Kassod tree	Caesalpiniaceae	June – Jan Peak in autumn	Yellow	A medium sized evergreen trees with drooping branches. It grows luxuriantly under humid conditions. It does not prefer water logging. Propagated by seeds. Grows vigorously both under sunny (or) semi-shady situation

69.	*Cassia spectabilis*	-	Caesalpin-iaceae	Sep - Dec	Golden yellow	An evergreen showy tree with spreading crown attaining a height of 10m. An ideal tree for planting along medium roads & in groups in parks. It is propagated by seeds
70.	*Castanospermum australe*	Moreton Bay chestnut	Caesalpin-iaceae	March	Yellow to orange	A large evergreen upright tree growing to a height of 15-18m. It is propagated from seeds. It is a shade tree for its dense foliage & handsome crown. It can be planted as avenue trees when wind is not strong, as its branches are weak& borne to breakage
71.	*Casuarina equisetifolia*	Beefwood/ Australian oak	Casuarin-iaceae	-	Pinkish	A tall, straight-stemmed tree reaching a height 0f 30m (or) more. It is mostly deciduous with drooping branches. It is a good tree in parks & large gardens. Due to its brittle nature of its branches, it is not advisable to plant them along the road sides. It is suitable for mass planting along the sea shore & river

						banks. It makes good hedge &topiary, if it is regularly pruned.
72.	*Cedrela microcarpa*	-	Meliaceae	-	White	A lofty tree with spreading branches. An evergreen tree of attractive appearance. It is suitable for roadside plantingIt is propagated from seeds
73.	*Cedrela toona*	Indian mahogany	Meliaceae	Feb - March	White fragrant	It is a rapidly growing, near evergreen tree, about 10m tall. It is a good roadside plant
74.	*Ceiba pentandra*	Kapok tree	Bombacaceae	Jan - March	Creamy white (or) light pink	A tall deciduous tree reaching 20m. It is suitable for growing in the saline tracts near the sea shore. It is a good tree for planting along wide roads, parks & in large gardens particularly along the border of large lawns. It can be planted behind a building to serve as a background
75.	*Cerbera manghas*	-	Apocynaceae	Throughout the year but profusely during summer	White with pink centre, fragrant	A small tree about 6m high. This small tree can be planted in groups in large gardens or at the back of the

						shrubbery. It is propagated by seeds (or) cuttings. It can stand salinity
76.	*Cerbera tanghin*	-	Apocynaceae	-	White	A dwarf evergreen with latex. It thrives in semi-shade & moist location. It can be planted on the banks of lake or tank
77.	*Chorisia speciosa*	Floss silk tree	Bombacaceae	October	Pinkish outer surface, lower half of inner surface pinkish-white gradually fading	A beautiful fast growing deciduous tree with bottle shaped trunk studded with large spines. Propagated from seeds. It is suitable for warm & moist localities, particularly coastal regions. It also grows well in dry regions. It looks beautiful when planted in row around the edge of a large green space. It can be planted in groups in parks and gardens & along medium roads
78.	*Chrysophyllum cainito*	Star apple	Sapotaceae	-	Purplish white	An evergreen tree reaching about 15m. It is propagated by seeds. It is a good avenue tree. It can be planted in row along the border of parks and gardens

79.	*Citharexylum quadrangulare*	Zither-wood	Verbenaceae	March - May	White scented	A small, deciduous tree, 5-6m high with drooping branches. It is propagated by seed (or) large cuttingsIt can be planted in small gardens (or) at the back of shrubberies (or) in groups in medium & large gardens & parks
80.	*Clitoria arborea*	-	Caesalpin-iaceae	May - June	Shades of violet, centre deeper with red markings	An evergreen, large tree with well shaped crown. It is suitable for planting in large gardens. It can be propagated from cuttings (or) air – layering
81.	*Clusia rosea*	Fat pork tree	Guttiferae	Jan - March	White with pink tinge	A well shaped evergreen tree with spreading crown reaching a height of 6-8m
82.	*Cochlospermum religiosum*	Yellow silk otton	Cochlospermaceae		Feb - March	Brilliant yellow. A soft-wooded, small to medium sized deciduous tree reaching a height of 6-10m. It is a quick growing tree. It can be planted at the background of shrubbery (or) along the border of large lawn (or) beside small & medium roads. It lowers in

						two years when it is of the size of a shrub
83.	*Colvillea racemosa*	Colville's Glory	Caesalpin-iaceae	Sep - Oct	Orange	It is a large deciduous tree of 14 – 16m high. When not in flowers it closely resembles *Delonix regia.* It is propagated from seedIt prefers a moist (or) moderately dry climate. It is a good tree for large avenues & for parks. It can be used for group planting
84.	*Cordia sebestena*	Aloe wood	Boraginaceae	Throughout the year	Orange scarlet	It is small evergreen tree (or) large shrub, reaching a height of 5-8m. It is propagated from seeds. It is specially suited for small gardensIn parks & large gardens, it can be planted at the back of shrubberies (or) in groups in the corners of a large lawn
85.	*Couroupita guianensis*	Cannon ball tree	Lecythi daceae	Several times in a year	Outside a mixture of white, yellow & inside deep pink (or) crimson	A soft wooded deciduous tree. It grows well in warm humid climate. It is propagated from seeds. It can be planted in parks, around temples & also wide city roads

86.	*Crataeva religiosa*	Bengal quince/ Caper tree	Cappari daceae	Nov-Dec	White (or) cream coloured	A moderate sized, much branches, spreading, deciduous tree. Reaching a height of about 10 – 15m. Propagated from seed & also by root suckers. It is a slow growing plantIt grows better in dry climate. Suitable for planting in medium, large gardens as well as parks. It can also be used as an avenue tree along medium roads
87.	*Crataeva religiosa* 'Nurvala'	-	Cappari daceae	-	-	-
88.	*Crescentia cujete*	Begger's bowl	Bignoniaceae	Rainy season	Yellowish with purplish veins	A small tree of 6-10m height, wide spreading. The tree is planted in gardens & parts. Propagated from seeds
89.	*Cupressus sempervirens*	Italian cypress	Cupressaceae	-	Yellow	A hardy, handsome tree. It is suitable for planting in parks, roadside planting Propagated by seeds. It is suitable to wide range of soil & climate
90.	*Cycas circinalis*	Fern palm	Cycadaceae	-	-	A palm like evergreen tree. It produces cylindrical trunk with the sear of fallen

						leavesIt is an excellent pot plant. Propagated from seeds & offsets, It is often planted in the corner (or) centre of big lawns
91.	*Cycas revoluta*	-	Cycadaceae	-	-	-
92.	*Dalbergia lanceolaria*	-	Caesalpin-iaceae	-	-	A tall & graceful tree. It is often grown on roadsides & timber tree in forest
93.	*Dalbergia sissoo*	South Indian Rosewood	Caesalpin-iaceae	-	Yellowish white scented	A large deciduous tree attaining a height of 20m (or) more. It grows both in dry & humid regions. A valuable timber tree. Good for planting along large road. National highways, canal side & river banks. Suitable for planting in large parks
94.	*Delonix elata*	-	Caesalpin-iaceae	Aug - March	White but turning yellow before fading	An erect, unarmed tree growing to a height of about 15-20m. Easily propagated by seeds. Suitable for planting along roads, avenues & in large gardens
95.	*Delonix regia*	Flame of forest	Caesalpin-iaceae	April - Aug	Orange -scarlet	A large, deciduous quick growing, open - branched

						spreading tree, reaching up to 15-20m Cannot with stand water logging & frost. Propagated from seeds & large cuttings
96.	*Dillenia indica*	Elephant apple	Dilleniaceae	June - July	White, sweet scented	An evergreen tree of 6-8m tall. Easily propagated from seeds, cuttings. Suitable for planting in gardens, parks & as well as along small avenues
97.	*Dillenia pentagyna*	-	Dilleniaceae	Mar-April	Yellow	A large tree having spreading branches of 15m height. A beautiful roadside tree
98.	*Diospyros peregrina*	River ebony	Ebeneceae	April - May	White, scented	An evergreen tree of moderate size. It has low spreading branches with dense rounded crown
99.	*Dolichandrone spathacea*	-	Bignoniaceae	March - May	white	This is a slender – looking, lofty deciduous tree of 15-18m high. It is propagated by seeds. It is a good tree for avenue & gardens
100.	*Dubanga sonneratioides*	-	Sonneratiacae	-	White	A tall – growing tree. It is grown in gardens & road side. Propagated from seeds

101. *Erythrina variegata*	Indian coral tree	Caesalpin-iaceae	Feb - March	Red	A soft wooded, deciduous, quick growing medium sized spreading tree reaching a height of 8-12mIt grows under frost free condition both in dry &moist regions. It can't stand water logging. Easily propagated from seeds & large woody cuttings. It is grown as live poles for fencing
102. *Erythrina varigata var. alba*	-	Caesalpiniaceae		-	- -
103. *Erythirina variegata var. picta*	-	Caesalpiniaceae		-	- -
104. *Eucalyptus calophylla*	-	Myrtaceae	-	White (or) cream colour	Tall tree with dense foliage. It is of slow growing habit. Good shade tree for avenues & parks in warm humid area. It cannot tolerate frost (or) very dry atmosphere
105. *Eucalyptus citriodora*	Lemon scented eucalyptus	Myrtaceae	March - May	white	A tall, quick growing tree to a height of 40-50m. It is suitable for dry & moist regionIt can't withstand frost. Propagation by seeds. Suitable for gardens & parks.

					They can also be planted along roads & highways in single (or) multiple rows &for afforestation
106. *Eucalyptus globulus*	Tasmanian blue gum	Myrtaceae	Dec - May	-	It is gereally sensitive to extreme cold & sudden changes of weather. It can be planted in large gardens & parks
107. *Eucalyptus maculata*	-	Myrtaceae	April - May	-	It comes up well in moist localities. It is also drought resistant. It can be tolerate extreme of temperature. It is a tall tree grows beyond 30m. Propagated by seeds
108. *Ficus benghalensis*	Banyan tree	Moraceae	-	-	Very large, deciduous, spreading tree reaching a height of 30m. It prefers warm humid climate. Easily propagated from seeds. Large pole cuttings. It may be planted in large gardens, parts & also wide roads. It is a favourite tree near temple, bathing ghats, river banks & on large meadows

109. *Ficus bengalensis* 'Krishnae'	-	Moraceae	-	-	-
110. *Ficus benjamina*	Java fig	Moraceae	-	-	Large, graceful, broad headed evergreen tree reaching a height of 15-20m. It prefers humid climate & moist soil. Propagated from seeds, cuttings. Suitable for planting along highways, large avenues, large parks, traffic islands, in the courtyards of temples. Good pot plant for indoor decoration
111. *Ficus elastica*	Indian Rubber tree	Moraceae	-	-	A huge, handsome, glabrous, quick growing, spreading tree. It grows above 30mIt prefers moist soil& humid climate. It can be tolerate shade. Suitable for roadside planting
112. *Ficus hispida*	-	Moraceae	-	-	It is a small tree
113. *Ficus infectoria*	-	Moraceae	-	-	A handsome, quick growing, large spreading, deciduous tree reaching a height of 15-18m. It grows both in the dry & humid frost, free tracts. It

					is mainly used as shade tree in parks & gardens & along large road. Propagated by seeds, large limbs
114. *Ficus religiosa*	Pipal tree	Moraceae	-	-	A large, glabrous, spreading tree, 15-20m height. It is propagated by seed. It prefers humid climate. It is a good shade tree for planting along the highways, in parks, market places (or) in large courtyards of big buildings for shade
115. *Ficus rumphii*	-	Moraceae	-	-	It is a glabrous tree of moderate size. A well shaped shade treeSuitable for roadside planting & public gardens
116. *Filicium decipiens*	Fern-leaf tree	Sapindaceae	-	white	It is a slow-growing evergreen tree. It can be planted along small avenues in parks, gardens (or) other protected places
117. *Firmiana colorata*	Bonefire tree	Sterculiaceae	Mar - April	Scarlet	It is a tall tree with rather short branches that give the tree a narrow appearance

118. *Fugosia hakeifolia*	-	Malvaceae	June	Yellow, a red purple spot at the base	A glabrous dwarf tree grows up tp 5mIt is propagated by cuttings & layering
119. *Gardenia latifolia*	Cape jasmine	Rubiaceae	Summer	White, highly scented	A small deciduous tree growing to a height of 5-6m. It is propagated by cuttings & air layering. It may be planted in shrubbery (or) along the garden path
120. *Gliricidia sepium*	Mata-raton	Caesalpin-iaceae	-	Pink, white, purpe	A quick growing, deciduous tree reaching a height of 15-20m. It prefers warm, humid climate. Propagated by seed & large cuttingsIt is suitable for roadside planting & along the path of a park. Used as a shade tree for plantation crops
121. *Grevillea robusta*	Silver Oak	Proteaceae	Mar - April	Orange	A tall, narrow, graceful tree grows to a height of 20 m. It can stand drought. It is easily propagated from seeds. Well suitable for group planting in parks, large gardens, also along roads, avenues
122. *Guaiacum officinale*	Tree of Life	Zygophy-llaceae	-	Blue	An evergreen slow growing medium sized tree reaching

					6-10m in heightIt thrives better in arid climate Propagated by seeds. It can be grown along small avenues & garden paths in small & medium gardens
123. *Gustavia augusta*	Stink wood	Lecythi daceae	Mar - Sep	White (or) tinged pink fragrant	A small bushy tree of 6-8m height, branching from near the base. It prefers a warm humid climate. Propagated from seed, cutting & air-layering The flowers look somewhat like magnolia. It can be planted in home gardens & parks
124. *Haematoxylon campechianum*	Logwood/ Campeachy tree	Mimosaceae	-	Yellow scented	A small, slow growing spreading tree of 6-7m height with a crooked, buttressed. It thrives in a hot & rather dry climateIt is suitable for planting along narrow roads, parks & gardens. It is propagated from seeds
125. *Heritiera macrophylla*	-	Sterculiaceae	Mar - April	Pinkish rose	A large evergreen tree. Suitable for planting for large public gardens from seeds. Propagated from seeds.

126. *Heterophragma adenophyllum*	-	Bignoniaceae	June - Aug	Corolla is of yellowish brown tube, swollen at the top	A large evergreen tree reaching a height of 15-20m. It grows in the warm climate both under humid & dry conditions. Propagated y seeds. It is good for planting along wide roads & boundary of large gardens & parks
127. *Hibiscus tiliaceus*	Cork wood/ Yellow mallow tree	Malvaceae	Summer & rainy	Yellow with crimson eye	A round headed spreading treeIt is grown in large gardens. Propagated from seeds &cuttings
128. *Holarrhena antidysenterica*	Easter tree/ Ivory tree	Apocynaceae	-	White (or) creamy scented	A small deciduous tree of 6-7m. It grows both in dry & wet localities. It can be propagated from seeds. It is suitable on garden paths, small compounds & at the back of shrubbery
129. *Holarrhena floribunda*	-	Apocynaceae	-	-	-
130. *Hura crepitans*	Sandbox tree	Euphorbiaceae	-	-	It grows up to 30m high. Tree is suited to light loam soil. Abundant milk juice secreted from wounds of the tree is poisonous. Propagated by cuttings (or) by seeds

131. *Inga dulcis*	Manilla tamarind	Euphorb-iaceae	Jan - April	-	A large evergreen tree grows to a height of up to 15m tall. Sharpe spines arising near the base of the leaves. It is common in dry & humid regions of India. It is grown along roads, boundaries of large gardens & farms. It is a popular hedge plant & responds to pruning. Propagated from seeds (or) cuttings
132. *Ipomoea arborescens*	-	Convolvu-laceae	Summer	White	Erect quick growing tree, reaching a height of up to 8m. Propagated from seeds, large cuttings. It is not widely grown in gardening
133. *Ixora parviflora*	Torch tree	Rubiaceae	Mar - May	White, rarely pink, scented	A small glabrous much branched evergreen tree. It prefers comparatively dry situation & warm temper-ature. It stands humid condition. Propagated from suckersIt is used along narrow garden paths & at the back of shrubbery. It is good as pot plants

134. *Jacaranda filicifolia*	Fern leaf Jacaranda	Bignoniaceae	Feb - Mar	Bluish purple	It stands damp soil. It is propagated from seeds. It is suitable for private gardens, parks, for planting along roads
135. *Jacaranda mimosaefolia*	Mimosa – leaved jacaranda	Bignoniaceae	Mar - June	Bluish purple	A medium sized, graceful, deciduous tree reaching a height of 8-10m. It prefers moderate climate & thrives in well drained soil. It is propagated from seeds & semi-hard wooded cuttings. It is grown widely along medium-sized roads, private & public gardens
136. *Juniperus chinensis*	Chinese juniper	Cupressaceae	-	-	It is a densely growing prickling small tree (or) large shrub reaching 3-5m height. It is a slow growing plant. It is used around green dams (or) along the garden paths at regular interval for its symmetrical & beautiful look. Twigs are susceptible to mealy bug attack. It is propagated from seeds

137. *Kigelia pinnata*	Sausage tree/ cucumber tree	Bignoniaceae	Mar - May	Deep chocolate red, evil smelling	A moderate sized, evergreen, spreading tree grows to a height of about 12-15m. Fruits are hard, cylindrical, cucumber like. It is a good avenue tree
138. *Kleinhovia hospita*	Tree antigonon	Sterculiaceae	July - Sep	Rose coloured	A medium sized tree up to 15mIt prefers warm humid climate. Easily propagated from seeds. It is a good shade tree. It is planted along highways & avenues. It can also be planted in large gardens & parks for its dense shade
139. *Lagerstroemia parviflora*	-	Lythraceae	Summer	white	A tall tree attaining a height of 20 mA fairly quick growing tree. Suitable for planting along avenues & in large gardens. It grows well both in dry & humid climate. It is propagated from seeds
140. *Lagerstroemia speciosa*	Pride of India	Lythraceae	April - June	Mauve	A very showy, large, deciduous tree with globular crown reaching a height of 15-20m. It prefers warm, humid climate. It can stand

					water logging. Propagated by seeds
141. *Lagerstroemia speciosa* 'Magnifica'		-	Lythraceae	-	- -
142. *Lagerstroemia speciosa* 'Rosea'		-	Lythraceae	-	- -
143. *Lagerstroemia thorelli*	-	Lythraceae	June - Sep	Mauve & fading to almost	A dwarf & bushy tree, 6-8 m high. It is propagated from seeds. It is often planted on roadside & in small & large gardens
144. *Leucaena leucocephala*	Horse tamarind	Mimosaceae	Mar – April & again Aug - Nov	White	A medium sized quick growing tree reaching 8m (or) more. It requires warm temperature. It can grow both under dry & humid conditions. Propagated from seeds. It can be planted along small & medium roads as shade tree. Wood is brittle & is frequently damaged by storm
145. *Magnolia grandiflora*	Lily tree	Magnoliaceae	Summer	White, sweet - scented	It is an evergreen, pyramidal tree reaches up to 15-20m. It prefers cool, humid climate. It is propagated by cuttings & layering. It should be planted in a site protected from western side

146. *Magnolia pterocarpa*	-	Magnoliaceae	April - May	White, fragrant	It is a beautiful, evergreen tree reaching 10-15m in height with dense rounded crown. It prefers warm, humid climates. It should be planted in large gardens
147. *Mangifera indica*	Mango tree	Anacard-iaceae	-	-	-
148. *Markhamia platycalyx*	-	Bignoniaceae	Feb - Mar	Yellow with red lines	A tall evergreen tree up to 15m height. It is suitable for planting on roadsides & gardens
149. *Melaleuca leucadendron*	White bottle brush/Swamp tea tree	Myrtaceae	Hot & rainy seasons	whitish	An evergreen tree with straight trunk. It is propagated fro seeds & cuttings. It is a slow growing tree. It can tolerate salinity. It is a good plant around water garden, lake & at the back of shrubberies
150. *Melia azedarach*	Indian lilac/ China berry	Meliaceae	-	Purple in colour, scented	A medium sized, fast growing handsome decid-uous tree reaching about 8-12m in height. It tolerate cold Easily propagated from fresh seed

151. *Mesua ferrea*	Iron-wood tree	Guttiferae	Feb - May	White, fragrant	A medium sized, handsome, slow – growing, evergreen tree up to 15m. It requires warm, humid climate, semi-shade. It is propagated from seed
152. *Michelia champaca*	Yellow champa	Magnoliaceae	Summer & rains	Yellow (or) dull orange, strongly scented	It is an evergreen tree growing to a height of about 12-15m. It prefers moderate temperature, humid atmosphere
153. *Michelia champaca* 'Alba'	-	Magnoliaceae	-	-	-
154. *Millettia peguensis*	Jewel on a string	Caesalpin-iaceae	Mar - April	Red (or) purplish	A medium sized deciduous tree of about 10m & semi-drooping branches. It prefers warm, humid climate. Propagated from seeds. It is well, suited for town roads, large private gardens & parks
155. *Millettia thonningii*	-	Caesalpin-iaceae	Mar - April	Reddish violet	A medium sized tree about 6m high with spreading & drooping branches. Propagated from seeds & cuttings

156. *Milligtonia hortensis*	Tree jasmine/ Indian cork tree	Bignoniaceae	-	White with greenish tinge, fragrant	It is propagated from suckers & cuttings. It is a quick growing tree. It is extensively planted along the roads, gardens & parks
157. *Mimusops elengi*	Indian Medlar	Sapotaceae	Spring & throughout warm season	Off – white fragrant	A medium sized, handsome evergreen tree with regular globular crown reaching a height of 10-15m. It is easily propagated from seeds. It is suitable for road side planting
158. *Mimusops elengi* 'Variegata'	-	Sapotaceae	-	-	-
159. *Monodora grandiflora*	African orchid	Annonaceae	Nov - Dec	Green with brown margin	It requires warms, humid climate. It is propagated from seeds
160. *Muntingia calabura*	Chinese cherry	Tiliaceae	-	White	A small spreading tree, branches drooping & pubescent. It is an excellent ornamental tree for lawns in all types of gardens. Propagated from seeds & root suckers
161. *Murraya koenigii*	Curry leaf tree	Rutaceae	-	White not scented	It is a small tree. It has dense, shady crown. It is grown in gardens for attractive foliage

					& sherry fruits. It is propagated by seeds & suckers
162. *Napoleona imperialis*	-	Lecythidaceae	Feb - June	Brown, rose & yellowish	It is a small evergreen tree growing up to 5m in height. It thrives well in semi-shade. It is protected from direct sun throughout the day. Propagated from seeds.
163. *Nauclea orientalis*	-	Rubiaceae	Summer months	Fragrant	A small tree with long quadrangular branches growing to a height of about 8-10m. It is propagated from seeds, cuttings
164. *Nyctanthes arbor-tristis*	Coral jasmine/ Tree of sadness/ Pavalamalli	Oleaceae	Autum	Orange (or) Orange – red, scented	It is a large shrub (or) small tree reaching a height of about 5-7m. It is easily propagated from seeds. It is planted in home garden for religious purpose
165. *Oroxylum indicum*	Indian trumpet flower	Bignoniaceae	June - July	Lurid - purple	A deciduous soft wooded tree grows up to 12m. It produces branches near the top. It is often grown along roads & also in gardens. Propagated from seeds & cuttings.

166. *Pachira cyathophora*	-	Bombacaceae	Mar - May	Cream - white	A tall evergreen tree which grows up to 15m (or) more. It produces semi-pendulous branches. It is propagated by seeds & cuttings
167. *Parkia biglandulosa*	Ball batmitton tree	Mimosaceae	Nov - Dec	Brown & velvety but becomes white as the flowers open	A large, upright, quick growing evergreen tree growing to a height of 20m (or) more. It thrives in warm & moist climate
168. *Parkinsonia aculeata*	Jerusalem thorn	Caesalpin-iaceae	Summer months	Yellow	A low tree growing to a height of 4-6m. It is easily propagated from seeds. It can be used as a hedge. It is particularly planted in the arid regions for soil conservation purpose
169. *Parmentiera cereifera*	Candle tree	Bignoniaceae	May - Aug	White (or) dull pink	A small tree, sometimes branching near the base. Fruit closely resembling old - fashioned tallow candle. It is a tree of tropical climate. It is easily propagated from seeds & cuttings. It is planted in gardens & parks
170. *Parmentiera edulis*	Guajilote tree	Bignoniaceae	-	Greenish yellow	It is a tall evergreen tree grows to height of 15m. It is

					planted in large gardens. It is propagated from seeds
171. *Peltophorum africanum*	-	Caesalpin-iaceae	Light yellow	June - Aug	It is slow growing. It produ-ces many branches from the base of the trunk. It can be effectively grown along roadsides & in gardens. It is easily propagated from seeds
172. *Peltophorum brasiliense*	-	Caesalpin-iaceae	-	Yellow	It is a graceful tree growing to a height of 15m with drooping branches. It is an ideal tree for a large, medium gardens
173. *Peltophorum pterocarpum*	Rusty shield bearer	Caesalpin-iaceae	Mar – June & also in Sep - Nov	Yellow	It is particularly deciduous tree growing to height of 20m. It grows equally well both in arid & moist localities. It is easily propagated from seeds & cuttings. It is largely planted along wide roads & in large gardens & parks. It can be alternated with *Delonix regia* which produce contrast red colour during the same flowering season
174. *Pinus longifolia*	Long-leaved pine	Pinaceae	-	-	It is a large subtropical tree growing to height of 20m. It

					can be planted as specimen tree in the centre of a big lawn. It makes fine avenue plants along the hilly roads. It is a good pot plant & for making bonsai when it is in young stage
175. *Piscidia erythrina*	Jamaica dog-wood	Caesalpin-iaceae	April	Summer	It is an excellent flowering tree grows to a height of 8-10m. It is suitable for planting on roadside & gardens
176. *Plumeria alba*	Pagoda tree	Apocynaceae	-	White (or) various shades of pink, yellow, red (scented)	It is a popular & free flowering tree. It is propagated from cuttings
177. *Plumeria lutea*	-	Apocynaceae	-	-	-
178. *Plumeria obtusa*	-	Apocynaceae	-	-	-
179. *Plumeria rubra*	-	Apocynaceae	-	-	-
180. *Plumeria rubra* forma bicolor	-	Apocynaceae	-	-	-
181. *Plumeria rubra* forma tricolor	-	Apocynaceae	-	-	-
182. *Plumeria rubra* forma typica	-	Apocynaceae	-	-	-

183. *Podocarpus elongatus*	African yellow wood	Podocarp-aceae	-	-	It is large, evergreen, resinous tree growing to a height of 20 – 22m
184. *Podocarpus macrophyllus*	Buddhist pine	Podacarp-aceae	-	-	An evergreen resinous tree growing to a height of up to 16m. It grows well in warm & humid tropics in partial shade. It is propagated from seeds (or) cuttings. It is a good pot plant
185. *Podocarpus neriifolius*	-	Podacar-paceae	-	-	-
186. *Polyalthia coffeoides*	-	Annonaceae	-	Greenish yellow	A well, shaped evergreen dwarf tree growing to a height of 8m
187. *Polyalthia longifolia*	Mast tree/ Indian fir	Annonaceae	Mar - April	Greenish yellow	A tall, deciduous handsome tree of pyramidal shapeIt can stand drought & hot climate. It stand water logging. It is favourite as avenue plants. They can planted in close rows along the boundary wall of a larger compound. Propagated from seeds

188. *Polyalthia longifolia* 'Angustifolia'	-	Annonaceae	-	-	-
189. *Polyalthia longifolia* 'Pendula'	-	Annonaceae	-	-	-
190. *Polyalthia suberosa*	-	Annonaceae	-	Yellow (or) Reddish brown	A small tree (or) tall shrub propagated from seeds.
191. *Pongamia pinnata*	Indian beech/ Poonga oil plant	Caesalpin-iaceae	-	Rose (or) liliac	A moderate sized, spreading, deciduous tree reaching to height of 10-12m. It is suitable for planting along the roads & highways, banks of canals & pools, parks & large gardens
192. *Prosopis julifora*	Algaoba bean/ Mesquite	Mimosaceae	April - May	Greenish turning yellow, fragrant	A deciduous tree reaching a height of up to 10m. It is also grown in gardens & parks. It is propagated by seeds. It is a plant of arid zone
193. *Pterocarpus indicus*	Burmese rose-wood/Padauk	Caesalpin-iaceae	-	Golden yellow fragrant	A semi-deciduous tree growing to a height of 20m (or) more. Propagated from seed (or) cuttings. It is a good tree for large gardens & parks. It is suitable for avenue plantation

194. *Pterospermum acerifolium*	Maple twist/ Muchkunda	Sterculiaceae	Spring & early summer	Pure white, scented	It prefers warm climate & high rainfall. It is easily propagated from seeds. It is a good tree on medium roads & in large gardens
195. *Pterospermum semisagittatum*	-	Sterculiaceae	-	White	It is a medium sized deciduous tree reaching a height of up to 15m. It is planted in large gardens. It is a beautiful tree with showy foliage
196. *Pterospermum suberifolium*	-	Sterculiaceae	Summer	Yellowish white scented	It is a handsome tree with spreading crown. It is a medium sized evergreen tree
197. *Pterygota alata*	Buddha's coconut	Sterculiaceae	Feb - March	Greenish purple (or) rusty brown outside & red veined within	It is a tall, deciduous tree reaching a height of up to 20m. It is grown on roadside & in large parks & other public gardens. It is a quick growing tree
198. *Pterygota alata* 'Diversifolia'	-	Sterculiaceae	-	-	-
199. *Putranjiva roxburghii*	Child-life tree/ Lucky bean tree	Sterculiaceae	Mar - May	Yellow	A medium sized evergreen tree growing to height of 10-15m. It is propagated from seeds. It is a good avenue tree for medium, sized town

						roadsAlso suitable for growing in gardens & parks in rows for their globular & shiny crown
200.	*Ravenella madagascariensis*	Traveller's palm	Musaceae	Rainy season	-	It prefer warm, humid climate. Propagated by suckers (or) seeds. It is highly suitable to plant in the middle of a lawn as an accent
201.	*Salix babylonica*	-	Saliaceae	-	-	A broad headed, medium sized deciduous tree reaching 10-12m height. It grows best in cool, subtropical climate & prefer moist soil. It is well suited for planting along canal (or) around water surface like the periphery of lakes, lily pool, water gardens etc. It is easily propagated by cuttings
202.	*Samanea samanSyn: Enterolobium saman*	Rain tree	Mimosaceae	April - Oct	Yellow	It is a very large, flat-topped deciduous tree reaching a height of 20-25m. It prefers warm, humid climate. It is easily propagated from seeds & also from large cuttings. It is very good for large, wide

					roads, highways & also for big parks
203. *Santalum album*	Sandal tree	Santalaceae	-	Purple	It is a small, evergreen tree reaching a height of 6m. It is easily propagated by seeds. It prefers well drained loam soil rich in humus. It is grown along the garden paths & at the back of shrubbery
204. *Sapium sebiferum*	Chinese tallow tree	Euphorb-iaceae	April – June & Oct - Nov	-	It is an evergreen tree of 6-8m height. It is a well shaped tree with spreading branches. It is an excellent tree for growing in moist soil & near rivers & streams
205. *Saraca asoca*	Asoka tree	Caesalpinia-ceae	Mar - May	Orange red	A handsome medium sized, spreading evergreen tree with rounded crown growing to height of 6-8m. It is hardy tree, grows well in warm, humid climate of tropical & sub tropical regions. It needs shelter from dry wind. It prefers semi-shady situationIt is very suitable for all gardens,

					parks. Easily propagated from seeds
206. *Saraca cauliflora*	-	Caesalpin-iaceae	Mar - May	Yellow & May	A much branched tree growing to a height of 8m tall. It is natives of India
207. *Saraca declinata*	Red saraca	Caesalpin-iaceae	Late Feb - April	-	A well shaped tree growing to a height of 8m. It is grown in tropical garden. It prefers partial sun & good humidity
208. *Saraca thaipingensis*	Yellow saraca	Caesalpin-iaceae	Jan - Mar	Golden yellow	It prefers partial sun, hot, dry climate, moist & rich soil It may be planted along medium roads, small & large gardens. Propagated from seeds & air-layering
209. *Schima wallichii*	-	Theaceae	May - July	White, fragrant	Trees grows up to 25m high. It grows well in tropical & mild subtropical climate. Often grown in foot hills. Propagated by seeds.
210. *Schleichera oleosa*	Lac tree/ Ceylon oak	Sapindaceae	Spring	Yellow	It is a large, deciduous tree grow up to 20m high. It is resistant to dry climate. It is a slow growing tree. It is a good tree for roadside planting, especially in dry

					area Propagated from seeds (or) root suckers. The lac cultured on this tree is the best in quality
211. *Schotia latifolia*	Elephant hedge bean tree	Caesalpin-iaceae	Summer	Dark crimson	A medium sized, unarmed tree growing to a height of bout 8-10m high. In young stage it is used as pot plant. It can be grown around lawns (or) along the garden paths. It is propagated by seeds
212. *Sesbania grandiflora*	Swamp pea	Caesalpin-iaceae	Sep - Nov	White, red, pink	It is a quick growing short lived tree, growing to height of 6m. It is propagated from seeds. It is mainly grown in rural areas for its flowers & pods which are also used as vegetables
213. *Spathodea campanulata*	African tulip tree/Scarlet bell tree	Bignoniaceae	Jan - Feb	Orange crimson outside, having yellow wrinkled margin	A handsome evergreen (or) deciduous tree growing to a height of about 15 – 20m. It is easily propagated from seeds, root suckers & cuttings. Suitable for medium & large gardens & avenues
214. *Sterculia villosa*	Dung tree	Sterculiaceae	-	Yellow with purple centre	A quick growing tree of elegant. A tall, deciduous

					spreading tree. It is propagated from seeds. It is planted along roadside & in large public gardens
215. *Stereospermum kunthianum*	-	Bignoniaceae	April - June	Dark purple changing to pale purple	A tall, evergreen tree grows up to 15m. It is a well shaped tree suitable for planting along roads & in large gardens. Propagated from seeds, cuttings & air-layering
216. *Swietenia macrophylla*	Large leaved mahogany	Meliaceae	-	White	It is atall, evergreen tree growing up to a height of 25-30m. The tree is more hardy than the other species
217. *Swietenia mahagoni*	Spanish mahogany	Meliaceae	Summer	Green	A large deciduous tree growing to a height of about 20-25m. It grows well in dry & moist climate. It is a slow growing tree. It is largely used as an avenue plants, parks & large gardens for shade. It is grown in forests for valuable timber
218. *Syzygium cumini*	Indian blackberry/ Java plum	Myrtaceae	Mar - June	White, fragrant	A large evergreen tree with many branches. A fairly quick growing hardy tree suitable for roadside

						planting in both arid & humid requires. Propagated by seeds & air-layering
219.	*Syzygium malaccense*	Malay rose-apple	Myrtaceae	Summer & Nov - Dec	Crimson	It is an evergreen, medium sized treeIt is planted in gardens for the showy. flowers & edible fruits Propagated by seeds.
220.	*Tabebuia argentea*	Golden bell	Bignoniaceae	Feb - April	Bright yellow	It is one of the very beautiful flowering tree in tropical & sub tropical regions. It is often damaged by strong wind. It is a very popular tree in small & large gardens (or) along roads & in the public places. Propagated from seeds & cuttings
221.	*Tabebuia avellandae*	-	Bignoniaceae	Winter	Reddish mauve	A deciduous tree growing to height of 10m width spreading branches. It can be grown on roadside, private gardens & parks & in other locations for beautification
222.	*Tabebuia chrysantha*	-	Bignoniaceae	Feb - Mar	Bright yellow	A small deciduous tree grows to a height of 5-6m. It can be grown on roadside, private gardens & parks & in

					other locations for beautification. An ideal tree for roadside planting & in small (or) large gardens
223. *Tabebuia donnell-smithii*	Gold tree	Bignoniaceae	Mar - May	Bright yellow	It is recommended for planting in both dry & humid regions
224. *Tabebuia pallida*	-	Bignoniaceae	Mar - May	Light pink & yellow outside	A quite growing tree in warm humid climate. Propagated from seeds & cuttings.
225. *Tabebuia palmeri*	-	Bignoniaceae	Nov - Jan	Lavender rose with white throat, striped yellow & purple	It grows to a height of up to 10m. AN extremely colourful tree for roadside, gardens. It is propagated from seeds & cuttings.
226. *Tabebuia rosea*	Rosy trumpet tree	Bignoniaceae	-	Rosy - pink	A tall erect deciduous tree grows to a height of 20m. It prefers warm & dry regions. It is a good roadside tree. Suitable for growing in parks & other public places.
227. *Tamarindus indica*	Tamarind	Caesalpiniaaceae		May - Aug	Red & yellow A large, handsome spreading usually deciduous tree grows up to 20-25m. It is propagated by seeds. It is a very hardy &

					strong avenue tree on wide roads & highways
228. *Tamarix articulata*	Tamarisk	Tamaricaceae	-	-	It grows well in dry, arid, windy tracts. It may be pruned after flowering to maintain shape & to encourage new growth. It is highly preferred in gardens & parks. It is suitable for dry & saline areas.
229. *Taxodium distichum*	Bald cypress/ Swamp cypress	Pinaceae	-	Purple	It is a tall deciduous tree usually pyramidal in shape. It is a graceful tree with attractive feathery foliage. Suitable for growing in swamps, rivers, lakes, parks & over calcareous rocks.
230. *Tectona grandis*	Teak	Verbenaceae	July - Aug	White	A large deciduous tree with tall, erect trunk reaching a height of 20-25m. It is a strong ornamental tree with large crown. Suitable for planting along roads,highways, boundary walls of factories, big compounds & parks.
231. *Terminalia arjuna*	White murdah	Combretaceae	April - May	Dull yellow	A large deciduous tree, grows to a height of 15-20m.

					It grows well in humid, tropical climate & prefers moist localities. It can be stand waterlogging. It is suitable for planting beside roads & highways & along the canals. Propagated by seeds.
232. *Terminalia catappa*	Indian almond	Combretaceae	Mar - April	White	It is a tall deciduous tree with branches which spread horizontally in whorls from a straight trunk. It grows specially near the coast, along roads & in parks.
233. *Thespesia populnea*	Indian tulip tree/ Portia tree	Malvaceae	April - May	Yellow & turn purplish with age	It is a small. Evergreen tree with umbrella like, dense hemispherical crown growing to a height of 8-10m. It is a quick growing tree. It is a good shade as well as flowering ornamental trees for planting along the road & private gardens & parks. It is easily propagated from seeds & cuttings. It can be stand salinity.
234. *Thevetia peruviana*	Yellow oleander	Apocynaceae	Throughout the year	Yellow	A small tree with milky juice. It flowers well in full

					sun & semi-shade. Propagated from seeds & cuttings
235. *Thuja orientalis*	Chine thuja	Cupressaceae	-	Bluish green	A small evergreen bushy tree (or) large compact shrub. It is highly attractive, dense foliage, with bushy habit of growth. It tolerates drought. Propagated from seeds & cuttings. It is suitable to grow in tropical, sub-tropical & warm – temperature regions.
236. *Tipuana tipu*	Pride of Bolivia/ Rosewood	Caesalpiniaceae	Several months from summer	Bright golden	A tall, semi-deciduous tree grows to a height of 15m (or) more. A well shaped tree with handsome foliage. Propagated from seeds. Suitable for tropical & sub tropical gardens.
237. *Vitis negundo*	Nochi	Lamiaceae	-	-	-
238. *Wrightia tomentosa*	-	Apocynaceae	April - July	Vermillion	A medium sized, deciduous tree with yellowish juice reaching up to 10m height. Suitable for small gardens. It can be planted along roads & paths. It prefers humid, tropical & sub tropical regions.

Chapter **5**

Shrubs

I. General information

1. **Shrub** is a woody or semi-woody perennial plant **with little or no trunk** and grows up to a height from 50 cm to less than four meters.
2. **Shrubbery border** is an area of the garden devoted exclusively for growing shrubs planted in a row of rectangular fashion.
3. Plants suitable for shrubbery border.

Scientific Name	Common Name	Notes
Bauhinia acuminata	Mandaram	Medium, white flowers
Bauhinia tomentosa	-	Medium to large, yellow flowers
Cassia biflorus	-	Dense, medium shrub, yellow flowers
Dombeya spectabilis	-	Medium tall, dense in growth, flowering in profusion
Duranta repens	-	A tall shrub, stands shearing
Hibiscus malvaviscus	-	Small to medium
Ixora arborea	Thechi	A small, spreading shrub
Malpighia coccigera	-	Woody, medium
Murraya exotica	Kamini	Hardy shrub, adapted to diverse conditions
Mussaenda erythrophylla	Mussanda	Medium, showy
Mussaenda philippica	-	Medium
Tabernaemontana coronaria	Nanthyavattam	Medium tall, woody

4. **Hedge** is nothing but planting of shrubs at regular intervals to form a thick screen.

5. **Tall growing shrubs** are *Azalea* spp., *Abutilon indicum, Buddleia asiatica, Buddleia madagascariensis, Brunfelsia calycina, Cassia glauca, Caesalpinia pulcherrima, Doembeya spectabilis, Ervatamia divaricata, Ficus panda, Fuchsia hybrida, Gardenia jasminoides, Pachystachys* spp., *Hamelia patens, Hamiltonia suaveolens, Hibiscus rosa sinensis, Hibiscus mutabilis, Holmskioldia sanguinea, Ixora parviflora, Jatropha multifida, Lagerstroemia indica, Murraya paniculata, Nyctanthes arbor-tristis, Nerium indicum, Poinsettia pulcherrima, Phyllanthus nivosus, Stenolobium stans,Thevetia peruviana, Tithonia tagetiflora, Weigela florida, Spartium junceum, Schefflera* spp., *Reinwardtia indica, Paeonia suffruiticosa, Osyris orborea, Ligustrum ovalifolium, Jocobinia carnea, Gmelina asiatica, Exocaria bicolor, Cassia glauca.*

6. **Shrubs of medium height** are *Acalypha* spp., *Achania malvaviscus, Cestrum nocturnum, Cestrum diurnum, Jasminum multiflorum, Lantana* spp., *Mussaenda luteola, Stachytarpheta indica, Tecoma capensis, Thunbergia erecta.*

7. **Shrubs of dwarf height** are *Barleria cristata, Daedalacanthus nervosus, Galphimiagracilis, Jasminum sambac, Lantana sellowiana, Plumbago auriculata, Russelia juncea, Russelia floribunda.*

8. **White flowering shrubs** are *Jasminum sambac, Cestrum diurnum, Jasminum multiflorum, Buddleia asiatica, Ervatamia divaricata, Murraya paniculata, Poinsettia pulcherrima, Nerium indicum, Nyctanthes arbor-tristis, Lantana camara, Ixora parviflora.*

9. **Blue flowering shrubs** are *Plumbago auriculata, Brunfelsia calycina, Daedalacanthus nervosus, Duranta plumieri.*

10. **Yellow flowering shrubs** are *Galphimiagracilis, Mussaenda luteola, Tecoma stans, Abutilon indicum, Spartium junceum, Reinwardtia indica, Gmelina asiatica, Excoecaria bicolor, Cassia glauca, Thevitia nerifolia.*

11. **Red flowering shrubs** are *Russeliajuncea, Achaniamalvaviscus, Lantana camara, Tecoma capensis, Nerium indicum, Lagerstroemia indica, Caesalpinia pulcherrima, Hamelia patens, Holmskioldia sanguinea, Jatropha multifida, Thevetia peruviana.*

12. **Shrubs of Indian origin** are *Jasminum sambac, Jasminumhumile, Ixoracoccinea, Nyctanthesarbor-tristis, Holmskioldia sanguinea, Clerodendruminerme, Ervatamiacoronaria, Crossandrainfundibuliformis, Plumbagorosea, Barleriacristata, Russeliajuncea, Jasminum multiflorum, Daedalacanthus nervosus.*

13. Common names for shrubs are as follows :

Abutilon megapotamicum 'Variegata'	:	Weeping Chinese lantern
Acalypha hispida	:	Cat's tail
Alternanthera amoena	:	Red parrot leaf
Alternanthera bettzickiana	:	Red calico plant
Alternanthera bettzickiana 'Aurea'	:	Yellow calico plant
Alternanthera versicolour	:	Copper leaf
Angelonia grandiflora	:	Gorgon flower
Barleria cristata	:	Philippine violet
Bauhinia acuminata	:	Orchid tree
Bauhinia galpini	:	Pride of the cape
Beloperone guttata	:	Shrimp plant
Brunfelsia americana	:	Lady of the Night
Brunfelsia calycina	:	Yesterday, today and tomorrow
Brunfelsia pauciflora	:	Yesterday, today and tomorrow
Brya ebenus	:	Jamaica ebony or green ebony
Buddleia asiatica	:	Summer lilac, Butterfly bush
Caesalpinia pulcherrima	:	Peacock flower, pride of Barbados
Cassia alata	:	Candle bush
Catasbaea spinosa	:	Prickly apple
Cestrum nocturnum	:	Raat ki Rani, Night queen
Cestrum diurnum	:	Din ka Raja, Day queen
Clerodendrum inerme	:	Forest jasmine
Clerodendrum paniculatum	:	Kashmir bouquet
Codiaeum variegatum	:	Croton, South sea laurel

Crossandra infundibuliformis	:	Fire cracker plant, Kanakambaram
Duranta plumeri	:	Pigeon berry, golden dew drop
Ervatamia divaricata	:	Chandni, Malti
Euonymus japonicas	:	Privet, Japanese Spindle
Euphorbia fulgens	:	Scarlet plume
Euphorbia heterophylla	:	Annual poinsettia
Euphorbia leucocephala	:	White poinsettia
Euphorbia pulcherrima		Poinsettia, Christmas flower
Euphorbia pulcherrima plenissima	:	Double poinsettia
Gardenia jasminoides	:	Cape jasmine
Gardenia radicans	:	Miniature gardenia
Hibiscus mutabilis	:	Changeable rose
Hibiscus rosa sinensis	:	China shoe, Gurhal, china rose, Semparuthi
Hibiscus schizopetalus	:	Coral hibiscus
Hibiscus syriacus	:	Rose of Sharon
Holmskioldia sanguinea	:	Cup and saucer
Iresine herbstii	:	Beefsteak plant
Iresine herbstii 'Acuminata'	:	Painted blood leaf
Iresine herbstii 'Aureo- reticulata'	:	Chicken gizzard
Iresine lindenii	:	Blood leaf
Ixora spp.	:	Rukmini, Idlipoo
Ixora barbata	:	Bearded ixora
Ixora chinensis	:	Chinese ixora
Jasminum sambac	:	Bela, Motia, Mogra, Gundumalli
Jasminum multiflorum	:	Kund, Kakada

Jasminum humile	:	Yellow jasmine
Jacobinia carnea	:	Flamingo plant
Lagerstroemia indica	:	Shravani, Crape myrtle
Murraya paniculata	:	Kamini
Murraya exotica	:	Chinese box
Nerium indicum	:	Lal Kaner, Arali
Nyctanthes arbor-tristis	:	Har Shingar, Tree of Sadness, Night Jasmine, Pavalamalli
Phyllanthus nivosus	:	Snow bush
Plumbago auriculata	:	Chitra, Cape leadwort
Russeliajuncea	:	Weeping Merry
Sambucus canadensis	:	American elder
Solanum macranthum	:	Brazilian potato tree
Tabernaemontana coronaria	:	Cape jasmine/Nandiavattam
Tecoma stans	:	Yellow bell
Tecomaria capensis	:	Cape-honeysuckle
Thevetia peruviana	:	Pili Kaner
Turnera ulmifolia	:	Sage rose

14. A shrub which is used as cut greens is *Breynia nivosa*
15. A shrub which is tolerant to acidic condition is *Azalea mollis*
16. *Hydrangea macrophylla* produces blue flowers in acidic soil
17. *Hydrangea macrophylla* produces pink/ red flowers in alkaline soil

18. Shrubs which belongs to the family **Acanthaceae** are

S. No.	Botanical Name	Common name	Flower colour	Flowering season
1.	*Aphelandra fascinator*	-	Orange - scarlet	-
2.	*Aphelandra fulgens*	-	Crimson	-
3.	*Aphelandra nitens*	-	Vermilion, scarlet	Summer
4.	*Aphelandra sinclairiana*	Coral aphelandra	Orange red	Summer
5.	*Aphelandra tetragona*	-	Bright scarlet	-
6.	*Asystasia chelonides*	-	Bluish purple	Summer & Rains
7.	*Asystasia gangetica*	-	Light purple	-
8.	*Barleria cristata*	Philippine viloet	-	-
9.	*Barleria prionites*	-	Yellow	Hot months
10.	*Barleria strigosa*	-	-	-
11.	*Barleria tomentosa*	-	Purplish blue	-
12.	*Beloperone amherstae*	-	Rose pink	-
13.	*Beloperone gutata*	Shrimp plant	Pinkish brown bracts & white flowers	Summer & rainy
14.	*Crossandra undulaefoliaSyn: C. infundibuliformis*	Fire craker plant	Orange yellow, deep orange, salmon pink	Throughout the year
15.	*Daedalacanthus macrophyllus*	-	Pale violet - blue	Winter & early summer
16.	*Daedalacanthus nervosus*	-	blue	Dec - Mar
17.	*Eranthemum alo-marginatum*	-	-	-
18.	*Eranthemum bicolor*	-	Dark red	-

19.	*Eranthemum laxiflorum*	-	Purple - rose	Year round flowering abundantly in rainy season
20.	*Eranthemum nigrum*	-	White with rosy centre	-
21.	*Eranthemum tricolor*	-	Rosy red	-
22.	*Graptophyllum pictum*	Caricature plant	Deep red	-
23.	*Jacobinia carnea*	Flamingo plant	pink	Throughout the year
24.	*Jacobinia gheisbreghtiana*	-	Scarlet	Summer
25.	*Jacobinia pauciflora*	-	Scarlet with yellow tips	Throughout the year
26.	*Justicia aurea*	-	Yellow	Summer
27.	*Justicia gendarussa*	-	Purple spots	summer
28.	*Justicia ovata*	-	Orange - scarlet	Many months except in winter
29.	*Pachystachys lutea*	Lollipop plant	White, emerging through golden yellow, showy bract	-
30.	*Pseuderanthemum alatum*	Chocolate plant	-	-
31.	*Pseuderanthemum atropurpureum*	-	Purpling with red marking, small size	-
32.	*Pseuderanthemum reticulatum*	-	-	-
33.	*Pseuderanthemum sinuatum*	-	White, spotted rosy	-
34.	*Ruellia lorenziana*	-	Bluish purple	Summer
35.	*Ruellia rosea*	-	Bright red	Throughout the year
36.	*Sanchezia nobilis*	-	Yellow, large	Summer & rainy

37.	*Strobilanthes dyerianus*	-	Violet – blue, funnel shaped	-
38.	*Strobilanthes glomeratus*	-	Bluish purple, funnel shaped	Summer & rainy
39.	*Thunbergia erecta*	-	Dark violet, orange throat	Throughout the year profusely during hot months
40.	*Thunbergia hybrida*	-	Blue-violet	Summer & rainy
41.	*Thunbergia natalensis*	-	Pale blue	Summer & rainy

19. Shrubs which belongs to the family **Amaranthaceae** are

S. No.	Botanical Name	Common name	Flower colour	Flowering season
1.	*Alternanthera amoena*	Red parrot leaf	-	-
2.	*Alternanthera bettzickiana*	Red calico plant	-	-
3.	*Alternanthera denatata*	-	-	-
4.	*Alternanthera versicolor*	Copper leaf	-	-
5.	*Iresine herbstii*	Beefsteak plant	-	-
6.	*Iresine lindenii*	Blood leaf	-	-

20. Shrubs which belongs to the family **Apocynaceae** are

S. No.	Botanical Name	Common name	Flower colour	Flowering season
1.	*Acokanthera spectabilis*	Winter green	White, fragrant	summer
2.	*Allamanda neriifolia*	-	Golden yellow	Summer & rains

3.	*Allamanda schottii*	-	Rich yellow, throat darker	Many months except in winter
4.	*Kopsia fruticosa*	-	Reddish white & tip and centre of the flower is red	Throughout the year
5.	*Nerium oleander*	Oleander	Pink, white, carmine	Summer & rainy
6.	*Pootia grandiflora*	-	White (or) creamy, scented	Summer & rainy
7.	*Stemmadenia bella*	-	White, large, scented yellow inside	Summer & early rains
8.	*Strophanthus dichotomous*	-	White	Summer & rainy
9.	*Strophanthus wallichii*	-	Yellow	Summer & rainy
10.	*Tabernaemontana coronaria*	Cape jasmine	White	Year round except in cold months
11.	*Tabernaemontana dichotoma*	Eve's apple	White scented	Hot months
12.	*Vinca major*	Madagascar periwinkle	-	-
13.	*Vinca minor*	-	-	-
14.	*Vinca rosea*	-	Rose (or) white	Year round flower

21. Shrubs which belongs to the family **Asclepiadaceae** are

S. No.	**Botanical Name**	**Common name**	**Flower colour**	**Flowering season**
1.	*Asclepias curassavicaSyn: A. fruticosus*	Blood flower/Ballon plant	Orange - red	Summer & rains
2.	*Asclepias physocarpa*	-	White	Summer

22. Shrubs which belongs to the family **Caesalpiniaceae** are

S. No.	Botanical Name	Common name	Flower colour	Flowering season
1.	*Bauhinia acuminata*	Orchid tree	Snow white	Mar - Sep
2.	*Bauhinia galpinii*	Pride of the cape	Brick red	-
3.	*Bauhinia petersiana*	-	White	-
4.	*Bauhinia tomentosa*	-	Sulphur – yellow with a dark brown blotch on the inner surface	Summer & rainy
5.	*Brya ebenus*	Jamaica ebony/Green ebony	Golden yellow	Throughout the year
6.	*Caesalpinia gillesii*	-	Yellow	Summer
7.	*Caesalpinia pulcherrima*	Pride of Barbados	Orange, scarlet	-
8.	*Cassia alata*	Candle bush	Corolla golden yellow	Winter months
9.	*Cassia bicapsularis*	-	Yellow	Summer & rainy
10.	*Cassia glauca*	-	Yellow	Throughout the year
11.	*Cassia laevigata*	-	yellow	-
12.	*Erythrina blakeii*	-	Deep red	Early summer to rainy season
13.	*Erythirna crista-galli*	-	Crimson - scarlet	Summer & rainy
14.	*Erythrina resupinata*	-	Bright scarlet	Feb - April
15.	*Ruttya fruticosa*	-	Orange red with a black blotch at the centre	Throughout the year

23. Shrubs which belongs to the family **Caprifoliaceae** are

S. No.	Botanical Name	Common name	Flower colour	Flowering season
1.	*Abelia chinensis*	-	White, fragrant	Summer
2.	*Abelia floribunda*	-	-	-
3.	*Abelia grandiflora*	-	White with a pinkish tinga	June - Nov

24. Shrubs which belongs to the family **Euphorbiaceae** are

S. No.	Botanical Name	Common name	Flower colour	Flowering season
1.	*Acalypha godseffiana*	-	-	Sunny situation
2.	*Acalypha hamiltoniana*	-	-	-
3.	*Acalypha hispida*	Cat's tail	-	-
4.	*Acalypha wilkesiana*	-	-	-
5.	*Codium variegatum*	Croton	-	-
6.	*Euphorbia cotinifolia*	Hierba mala	Yellow	-
7.	*Euphorbia fulgens*	Scarlet plume	Orange red	-
8.	*Euphorbia heterophylla*	Annual poinsettia	-	-
9.	*Euphorbia leucocephala*	White poinsettia	Small yellow	-
10.	*Euphorbia pulcherrima*	Chrismas flower	-	Winter months
11.	*Excoecaria bicolor*	-	Yellow, small	-
12.	*Jatropha panduraefolia*	-	red	-

13.	*Manihot esculenta*	-	-	-
14.	*Phyllanthus nivosus*	Snow bush	Greenish	-

25. Shrubs which belongs to the family **Loganiaceae** are

S. No.	**Botanical Name**	**Common name**	**Flower colour**	**Flowering season**
1.	*Buddleia asiatica*	-	White, scented	-
2.	*Buddleia davidii*	-	Lilac, fragrant	Summer & early rains
3.	*Buddleia lindleyana*	-	Purplish violet	-
4.	*Buddleia madagascariensis*	-	Orange yellow	Spring & summer

26. Shrubs which belongs to the family **Lythraceae** are

S. No.	**Botanical Name**	**Common name**	**Flower colour**	**Flowering season**
1.	*Cuphea hyssopifolia*	False heather	Purplish rose	-
2.	*Cuphea melvillea*	-	Red at base, yellow at open	-
3.	*Cuphea miniata*	-	Bright red	-

27. Shrubs which belongs to the family **Labiatae** are

S. No.	**Botanical Name**	**Common name**	**Flower colour**	**Flowering season**
1.	*Coleus blumei*	-	-	-
2.	*Coleus rehneltianus*	-	-	-

28. Shrubs which belongs to the family **Malvaceae** are

S. No.	Botanical Name	Common name	Flower colour	Flowering season
1.	*Abutilon darwinii*	Flowering maple	Orange – red with scarlet veining	June - Oct
2.	*Abutilon hybridum 'Variegatum'*	-	-	-
3.	*Abutilon megapotamicum 'Variegata'*	Weeping Chinese lantern	Yellow	Spring to Autumn
4.	*Abutilon striatum*	Spotted flowering maple	Orange with red veins	Several months except in winter
5.	*Hibiscus colllinus*	-	White with purplish brown centre	Summer & rainy
6.	*Hibiscus mutabilis*	Changeable rose	Petals white at first, changing to deep pink & finally red	Sep - Dec
7.	*Hibiscus rosa-sinensis*	China rosse	White, yellow, orange, red, pink, salmon & purple	Throughout the year
8.	*Hibiscus schizopetalus*	Coral hibiscus	Orange - red	Throughout the year
9.	*Hibiscus syriacus*	Rose of Sharon	Rose, blue, purple, white, violet	Throughout the year
10.	*Malvaviscus arboreus*	-	Scarlet	Year round

29. Shrubs which belongs to the family **Mimosaceae** are

S. No.	Botanical Name	Flower colour	Flowering season
1.	*Calliandra brevipes*	Pink stamens	Throughout the year
2.	*Calliandra emarginata*	Scarlet	Summer & rainy
3.	*Calliandra haematocephala*	Crimson	-
4.	*Calliandra houstonii*	Flower buds are dark brown, stamens are purple - white	Year round
5.	*Calliandra hybrida*	-	-
6.	*Calliandra protoricensis*	White	Winter & spring
7.	*Calliandra tweedii*	Upper part of stamens pink, lower part white	Throughout the year

30. Shrubs which belongs to the family **Rubiaceae** are

S. No.	Botanical Name	Common name	Flower colour	Flowering season
1.	*Canthium glabrum*	-	-	Summer
2.	*Catesbaea spinosa*	Prickly apple	Greenish white turning to yellow with age	summer
3.	*Gardenia jasminoides*	Cape Jasmine	White colour, gradually turning yellow, fragrant	Summer & early rainy

4.	*Gardenia longistyla*	-	White	Hot months
5.	*Gardenia lucida*	-	White, scented	Summer & rainy
6.	*Gardenia radicans*	Miniature cardenia	White	Summer
7.	*Gardenia thunbergia*	-	White, scented	Summer
8.	*Hamelia patens*	-	Orange scarlet	Year round
9.	*Hamiltonia suaveolens*	-	Pale heliotrope in colour	Winter
10.	*Ixora acuminata*	-	White, fragrant	-
11.	*Ixora aliporensis*	-	Orange	-
12.	*Ixora barbata*	Bearded ixora	white	-
13.	*Ixora chinensis*	Chinese ixora	Red	-
14.	*Ixora coccinea*	-	Deep scarlet	-
15.	*Ixora duffii*	-	Orange scarlet	Summer & rainy
16.	*Ixora fulgens*	-	Orange scarlet	-
17.	*Ixora hybrida*	-	-	-
18.	*Ixora lutea*	-	Yellow	-
19.	*Ixora macrothyrsa*	-	Bright scarlet	-
20.	*Ixora profuse*	—	Yellow	-
21.	*Ixora spectabilis*	-	Bright orange	-
22.	*Ixora superba*	-	White, scented	-
23.	*Ixora venusta*	-	Pale pink	-
24.	*Lindenia rivalis*	-	White, scented	Summer

25.	*Mussaenda erythrophylla*	-	Bright scarlet above whitish with red veins & below hairy on both surfaces	Summer & rainy
26.	*Mussaenda frondosa*	-	Orange, expanded single sepal, creamy white	Summer & rainy
27.	*Mussaenda luteola*	-	Yellow	April - Sep
28.	*Mussaenda philippica 'Aurorae'*	-	Yellow	April - Oct
29.	*Pentas lanceolata*	Egyptian star cluster	Star cluster	-
30.	*Pentas karmesiana*	-	-	-
31.	*Pogonopus exsertus*	-	Pink	Throughout the year
32.	*Portlandia grandiflora*	-	Large white, pinkish inside the throat	Summer & rainy
33.	*Posoqueria latifolia*	Needle flower	-	Mar – May
34.	*Randia macrantha*	-	White, large turning pale yellow	Summer
35.	*Randia maculate*	-	-	-
36.	*Rondeletia odorata*	-	*Vermillion with* conspicuous, yellow throat	*7-8 months in a year*
37.	*Rondeletia strigosa*	-	Bronzy crimson with orange throat	7-8 months in a year
38.	*Sarcocephalus cordatus*	-	White	April - May
39.	*Serissa foetida*	-	White, light pink beneath	Summer & rainy
40.	*Tarenna zeylancia*	-	White, highly scented	Throughout the year, profuse flowering in summer

31. Shrubs which belongs to the family **Scrophulariaceae** are

S. No.	Botanical Name	Common name	Flower colour	Flowering season
1.	*Angelonia angustifolia*	-	Deep violet	-
2.	*Angelonia grandiflora*	Gorgon flower	White with violet spots	Throughout the year
3.	*Russelia juncea*	Coral plant	Bright red	Summer & rainy
4.	*Russelia sarmentosa*	Coral blow	Red	-

32. Shrubs which belongs to the family **Solanaceae** are

S. No.	Botanical Name	Common name	Flower colour	Flowering season
1.	*Brunfelsia americana*	-	White, fading to yellow, sweet scented	-
2.	*Brunfelsia calycina*	Yesterday-today-tomorrow	First violet with a yellow centre changing to dull purple & fading to white	Summer & rainy
3.	*Brunfelsia undulata*	-	White, fading to yellowish	Summer & rainy
4.	*Cestrum aurantiacum*	-	Orange yellow	Several months
5.	*Cestrum diurnum*	Day Jasmine	White, scented	-
6.	*Cestrum elegans*	-	Reddish purple	-
7.	*Cestrum nocturnum*	Queen of night	Creamy white	Year round
8.	*Cestrum parqui*	Willoe-leaved jasmine	Yellow	Summer & Rainy

9.	*Datura chlorantha*	-	Yellow, fragrant	-
10.	*Datura fastuosa*	-	Violet outside, whitish within	-
11.	*Datura sanguine*	-	Orange red	-
12.	*Datura suaveolens*	Angel's trumpet	White, fragrant	Summer & Rainy
13.	*Iochroma tubulosum*	-	Violet	Year round
14.	*Solanum macranthum*	Brazilian potato tree	Bluish violet fading to white	Throughout the year
15.	*Solanum rantonnetti*	-	Violet with yellow centres	April – Aug
16.	*Streptosolen jamesonii*	-	Yellow turning reddish orange, funnel shaped	Throughout the year

33. Shrubs which belongs to the family **Verbenaceae** are

S. No.	Botanical Name	Common name	Flower colour	Flowering season
1.	*Callicarpa cana*	-	lilac	Summer & rainy
2.	*Caryopteris mastacanthus*	Blue spirea	Violet	-
3.	*Clerodendrum fragrans*	-	White, tinged with pink, fragrant	Summer & rainy
4.	*Clerodendrum inerme*	Forest jasmine	White	Rains
5.	*Clerodendrum macrosiphon*	-	white	-
6.	*Clerodendrum nutants*	-	White	Summer & rainy
7.	*Clerodendrum paniculatum*	Kashmir bouquet	Scarlet	Summer
8.	*Clerodendrum phlomidis*	-	White	Oct - Nov

9.	*Clerodendrum siphonanthus*	Tube flower	White turning cream	Later summer, early monsoon
10.	*Clerodendrum variegatum*	-	-	-
11.	*Duranta lorentzii*	-	White	-
12.	*Durantha repens*	Golden dew drop	Blue	-
13.	*Gmelina asiatica*	-	yellow	Summer & Rains
14.	*Lantana camara*	-	Colour of flower, often change with age	-
15.	*Lantana sellowiana*	-	Mauve on flat heads	Cold months
16.	*Petrea arborea*	-	Purplish blue	Summer & rainy
17.	*Stachytarpheta indica*	-	Blue on terminal slender spike	Summer & rainy
18.	*Stachytarpheta mutabilis*	-	Crimson, fading to rose	Summer & rainy
19.	*Vitex agnus-castus*	Chaste tree	Mauve blue (or) silver blue	Mild summer until autumn

II. Choose the best answer

1. Pink flowering shrub is................

   Jocobinia carnea *Abutilon indicum* *Spartium junceum*

2. Common name of *Bauhinia acuminata* is................

   Shrimp plant Orchid tree Copper leaf

3. Example for tall growing shrub is................

   Thunbergia erecta *Mussaenda luteola* *Caesalpinia pulcherrima*

4. *Clerodendruminerme* is native of................

   India Africa America

5. Flower colour of *Hydrangea macrophylla* producedin acidic soil is.........

   Pink Red Blue

III. Match the following

1.	*Russeliajuncea*	:	Yellow flower
2.	*Lagerstroemia indica*	:	Red flower
3.	Copper leaf	:	White flower
4.	*Cassia glauca*	:	Purple flower
5.	*Ixora parviflora*	:	Copper leaf

Answers

II. Choose the best answer

1. *Jocobinia carnea* 2. Orchid tree
3. *Caesalpinia pulcherrima* 4. India
5. Blue

III. Match the following

1. Red flower 2. Purple flower
3. Copper leaf 4. Yellow flower
5. White flower

ANNEXURE – II SHRUBS

S. No.	Botanical Name	Common name	Family	Flower colour	Flowering period	Remarks
1.	*Abelia chinensis*	-	Caprifoli-aceae	White, fragrant	Summer	-
2.	*Abelia floribunda*	-	Caprifoli-aceae	-	-	-
3.	*Abelia grandiflora*	-	Caprifoli-aceae	White with a pinkish tinga	June - Nov	It can be planted in full sun (or) partial shade. It responds to light pruning after flowering. Propagated by semi hard wood cuttings
4.	*Abutilon darwinii*	Flowering maple	Malvaceae	Orange – red with scarlet veining	June - Oct but pendulous	Flowers look like hibiscus
5.	*Abutilon hybridum 'Variegatum'*	-	Malvaceae	-	-	-
6.	*Abutilon megapotamicum 'Variegata'*	Weeping Chinese lantern	Malvaceae	Yellow	Spring to Autumn	-
7.	*Abutilon striatum*	Spotted flowering maple	Malvaceae	Orange with red veins	Several months except in winter	It flowers profusely in partial shade & full sun. Propagated by cuttings

8.	*Acalypha godseffiana*	-	Euphorbiaceae	-	Sunny situation	It is one of the colourful shrubs with drooping branches. Suitable for growing in pots. Pruning is necessary to develop to maintain good appearance Propagated from seeds.
9.	*Acalypha hamiltoniana*	-	Euphorbiaceae	-	-	It produces flowers on drooping spikes. Pruning is necessary to develop a bushy shrub. It suitable for full sunny condition. Water stagnation should be avoided. It is propagated by cuttings
10.	*Acalypha hispida*	Cat's tail	Euphorbiaceae	-	-	It is suitable for planting in shrubbery & also as hedge (or) for screening. It is used as pot plant. They do best in full sun It responds to pruning.
11.	*Acalypha wilkesiana*	-	Euphorbiaceae	-	-	
12.	*Acokanthera spectabilis*	Winter green	Apocynaceae	White, fragrant	summer	It is propagated by cuttings, air layering & seeds.
13.	*Allamanda neriifolia*	-	Apocynaceae	Golden yellow	Summer & rains	It is also well adopted for planting against a back wall

						(or) training on pillars (or) trellis. It is multiplied by cuttings (or) layers. Flower colour is smaller than Allamanda cathartica
14.	*Allamanda schottii*	-	Apocynaceae	Rich yellow, throat darker	Many months except in winter	A hardy plant, suitable for shrubbery (or) for training on support. It responds to pruning. Propagated by cutting
15.	*Alternanthera amoena*	Red parrot leaf	Amaranth-aceae	-	-	Leaves are of different colour like carmine, pink, purple & yellow
16.	*Alternanthera bettzickiana*	Red calico plant	Amaranth-aceae	-	-	Leaves are available in light yellow, pink & light purple colours
17.	*Alternanthera denatata*	-	Amaranth-aceae	-	-	Spreading, semi-woody plants with deep red branches. A showy plant for growing in ground & pot in full sun & semi-shade
18.	*Alternanthera versicolor*	Copper leaf	Amaranth-aceae	-	-	Leaves are coppery red in colour. They are also grown in shrubbery & in pots. It responds to pruning. Propagated by cuttings.

19.	*Angelonia angustifolia*	-	Scrophula-riaceae	Deep violet	-	A bushy, perennial shrub
20.	*Angelonia grandiflora*	Gorgon flower	Scrophula-riaceae	White with violet spots	Throughout the year	It prefers partial shade. Propagated by cuttings. It responds to pruning
21.	*Aphelandra fascinator*	-	Acanthaceae	Orange - scarlet	-	A dwarf shrub grows up to 70cm height
22.	*Aphelandra fulgens*	-	Acanthaceae	Crimson	-	A dwarf semi-woody shrub about 50 cm tall
23.	*Aphelandra nitens*	-	Acanthaceae	Vermilion, scarlet	Summer	A dwarf, compact, semi-woody shrubs
24.	*Aphelandra sinclairiana*	Coral aphelandra	Acanthaceae	Orange red	Summer	-
25.	*Aphelandra tetragona*	-	Acanthaceae	Bright scarlet	-	An evergreen erect shrub with spreading branches. It flowers in shady & semi-shady locations. Propagated by cuttings
26.	*Asclepias curassavica*	Blood flower	Asclepia daceae	Orange - red	Summer & rains	A herbaceous perennial shrub growing up to 1m highIt is commonly used in borders & for planting as specimen bush. It is a good pot plant. It is propagated by seeds (or) cuttings

27.	*Asclepias physocarpa*	-	Asclepiad-aceae	White	Summer	A slender growing shrub about 2m high with white hairs. It is grown on shrubbery & in pot. It is propagated from seeds (or) cuttings
28.	*Asystasia chelonides*	-	Acanthaceae	Bluish purple	Summer & Rains	A dwarf, semi woody shrub
29.	*Asystasia gangetica*	-	Acanthaceae	Light purple	-	They are commonly planted in the shrubbery both in the sun & shade. It is propagated by seeds & cuttings. A dwarf, semi-shrub grows to a height of 30 – 45cm
30.	*Barleria cristata*	Philippine viloet	Acanthaceae	-	-	A dwarf bushy, much – branched shrub grows to a height of 0.75-1.5m
31.	*Barleria prionites*	-	Acanthaceae	Yellow	Hot months	An evergreen, much branched, prickly bush grows to a height of 1.5m. It grows well in sun, semi-shade
32.	*Barleria strigosa*	-	Acanthaceae	-	-	A bushy shrub
33.	*Barleria tomentosa*	-	Acanthaceae	Purplish blue	-	It produces compact bushy growth. It responds pruning. It grows well in sun & partial shade. It is grow as

						hedge (or) in shrubbery. Propagated by tip cuttings
34.	*Bauhinia acuminata*	Orchid tree	Caesalpinia-ceae	Snow white	Mar - Sep	It is a hardy plant & grows widely in private gardens, parks & public places. It is propagated by seeds & cuttings. It responds to pruning
35.	*Bauhinia galpinii*	Pride of the cape	Caesalpini-aceae	Brick red	-	A bushy, rambling, prostrate shrubPropagated by seeds & layering
36.	*Bauhinia petersiana*	-	Caesalpini-aceae	White	-	It is a tall shrub with many branches reaching a height of 3m. It is a large spreading shrub
37.	*Bauhinia tomentosa*	-	Caesapini-aceae	Sulphur – yellow with a dark brown blotch on the inner surface	Summer &rainy	An erect, bushy, handsome deciduous shrub growing to a height of 2.5m. It responds to pruning. It is propagated by cuttings
38.	*Beloperone amherstae*	-	Acanthaceae	Rose pink	-	A quick growing dwarf bushy shrub, 50-80cm in height
39.	*Beloperone gutata*	Shrimp plant	Acanthaceae	Pinkish brown bracts & white flowers	Summer & rainy	A dwarf shrub growing to a height of about 60cm in high. It responds to pruning.

						It is widely grown in pots for indoor gardening.
40.	*Brunfelsia americana*	-	Solanaceae	White, fading to yellow, sweet scented	-	A medium sized, bushy shrub growing to a height of 2m
41.	*Brunfelsia calycina*	Yesterday-today-tomorrow	Solanaceae	First violet with a yellow centre changing to dull purple & fading to white	Summer & rainy	A well shaped bushy flowering shrub growing to a height of 2m. It is propagated by air-layering (or) tip cutting. It grows well in full sun & partial shade
42.	*Brunfelsa undulata*	-	Solanaceae	White, fading to yellowish	Summer & rainy	It prefers full sun & semi-shade. Propagated by cutting (or) air layering. It responds to light pruning
43.	*Brya ebenus*	Jamaica ebony/ Green ebony	Caesalpin-iaceae	Golden yellow	Throughout the year	It is a tall shrub growing to a height of up to 4m. It can be grown at the back of the shrubbery (or) as a specimen plant. It produces pendulous branches. Propagated by seeds, cuttings & air - layering
44.	*Buddleia asiatica*	-	Loganiaceae	White, scented	-	An evergreen tall shrub growing to a height of 3m

45.	*Buddleia davidii*	-	Loganiaceae	Lilac, fragrant	Summer & early rains	A large deciduous shrub
46.	*Buddleia lindleyana*	-	Loganiaceae	Purplish violet	-	A bushy spreading shrub growing to a height of about 2m
47.	*Buddleia madagascariensis*	-	Loganiaceae	Orange yellow	Spring & summer	An evergreen tall shrub growing to a height of about 2.5m. It grows well in full sun & partial shade. It responds to pruning. Propagated by cuttings
48.	*Caesalpinia gillesii*	-	Caesalpini-aceae	Yellow	Summer	A tall shrub. It prefers full sun. Propagated by seeds, suckers (or) layering
49.	*Caesalpinia pulcherrima*	Pride of Barbados	Caesalpini aceae-	Orange, scarlet	-	A bushy shrub with a few prickles growing to a height of about 2.5 to 3m
50.	*Calliandra brevipes*	-	Mimosaceae	Pink stamens	Throughout the year	A bushy shrub with drooping & spreading branches growing to a height of about 1.5 – 2m
51.	*Calliandra emarginata*	-	Mimosaceae	Scarlet	Summer & rainy	It prefers full sun. Propagated by seeds, air layering. It does not respond to pruning
52.	*Calliandra haematocephala*	-	Mimosaceae	Crimson	-	A tall spreading shrub growing to a height of about

No.	Botanical name	Common name	Family	Flower colour	Flowering season	Remarks
						2.5m. A well shaped ever green shrub. Propagated from seeds, cuttings & layers
53.	*Calliandra houstonii*	-	Mimosaceae	Flower buds are dark brown, stamens are purple - white	Year round	A perpetual flowering straggling plant with terete branches. It grows better in shady place. Propagated by cuttings & air - layering
54.	*Calliandra hybrida*	-	Mimosaceae	-	-	-
55.	*Calliandra protoricensis*	-	Mimosaceae	White	Winter & spring	A large, tall shrub growing to a height of about 2.5m. Propagated from seeds (or) cuttings. They are fairly quick growing, need pruning almost every year
56.	*Calliandra tweedii*	-	Mimosaceae	Upper part of stamens pink, lower part white	Throughout the year	A tall shrub growing to a height of about 3m. It is a hardy shrub suitable for growing in the shrubbery (or) as a specimen plant. It produces branches almost horizontally
57.	*Callicarpa cana*	-	Verbenaceae	lilac	Summer & rainy	A bushy tomentose plant growing to a height of 2.5m. A hardy & quick growing shrub. Propagated by cuttings

58.	*Cananga kirkii*	-	Annonaceae	Greenish yellow	Mar - Sep	A tall shrub reaching a height of 3m. It is propagated by cuttings (or) air - layering
59.	*Canthium glabrum*	-	Rubiaceae	-	Summer	A tall spreading evergreen shrub growing to a height of 3m. A hardy & quick growing habit. It responds to pruning
60.	*Capparis spinosa*	-	Cappari daceae	White petals, upper petals are tinged yellow & inside turning dark red	Early summer	A much branched woody shrub growing to a height of 2m. It grows well in partial shade. Propagated by cuttings
61.	*Caryopteris mastacanthus*	Blue spirea	Verbenaceae	Violet	-	A very hardy, deciduous shrub growing to a height of 1.8 – 2.4m. Propagated by cuttings
62.	*Cassia alata*	Candle bush	Caesalpini-aceae	Corolla golden yellow	Winter months	A quick growing shrub. It prefers sun & semi-shade. It is hardly eaten by cattle
63.	*Cassia bicapsularis*	-	Caesalpini-aceae	Yellow	Summer & rainy	It is a bushy shrub growing to a height of 2m with arching branches. It is a free flowering shrub. Propagated by seeds, cuttings (or) layers

64.	*Cassia glauca*	-	Caesalpiniaceae	Yellow	Throughout the year	A tall, bushy, evergreen plant growing to a height of 3m. Itis propagated by seeds
65.	*Cassia laevigata*	-	Caesalpiniaceae	yellow	-	A tall bushy shrub growing to height of up to 4m. It is a very hardy shrub. It prefers both full sun & partial shade
66.	*Catesbaea spinosa*	Prickly apple	Rubiaceae	Greenish white turning to yellow with age	summer	A thorny shrub growing to a height of up to 25m. It is a slow growing hardy plant
67.	*Cestrum aurantiacum*	-	Solanaceae	Orange yellow	Several months	An erect (or) scrambling shrub glabrous and pubescent when young. It forms a bushy plant. Branches droop on all sides. It can grow in partial shade
68.	*Cestrum diurnum*	Day Jasmine	Solanaceae	White, scented	-	A bushy shrub with erect & spreading branches. A quick growing evergreen shrub Pruning is recommended after flowering

69.	*Cestrum elegans*	-	Solanaceae	Reddish purple	-	A tall, rambling, evergreen shrub. It grows well in full sun & partial shade
70.	*Cestrum nocturnum*	Queen of night	Solanaceae	Creamy white	Year round	A large evergreen subscandent shrub. Flowers are highly scented at night. A quick growing shrub. It is a hardy shrub & suitable for screening & can be trained on a trellis & low walls
71.	*Cestrum parqui*	Willoe-leaved jasmine	Solanaceae	Yellow	Summer & Rainy	It is a tall, deciduous shrub. It is easily propagated by cuttings.
72.	*Clerodendrum fragrans*	-	Verbenaceae	White, tinged with pink, fragrant	Summer & rainy	An erect growing shrub. It is a hardy & free flowering shrub. It grows better in moist semi-shady places. It is require hardy & drought resistant. It is propagated by separation of root suckers & cuttings
73.	*Clerodendrum inerme*	Forest jasmine	Verbenanacae	White	Rains	A straggling, evergreen shrub. It can be used for topiary work. It is propagated by cuttings. It grows well in swampy

						areas. It stands salinity well. It is less damaged by cattle
74.	*Clerodendrum macrosiphon*	-	Verbenaceae	white	-	A beautiful, dwarf, bushy shrub grows to a height of 1m. It is a free, intermittent flowering plant. It can be grown in the front row of a shrubbery. Propagated by tip cuttings
75.	*Clerodendrum nutants*	-	Verbenaceae	White	Summer & rainy	It is a very hardy shrub. It grows well in semi-shade. Propagated by cuttings (or) seeds.
76.	*Clerodendrum paniculatum*	Kashmir bouquet	Verbenaceae	Scarlet	Summer	It is a tall shrub growing to a height of 2.5m. It grows well in full & partial sun. It is propagated by suckers & cuttings. It is recommended for pruning after flowering
77.	*Clerodendrum phlomidis*	-	Verbenaceae	White	Oct - Nov	It is a large bush. It maintains bushy appearance. Propagated by cuttings
78.	*Clerodendrum siphonanthus*	Tube flower	Verbenaceae	White turning cream	Later summer, early monsoon	A tall shrub growing to a height of 2.5m. It grows well in full sun & partial shade. It stands ill drainage. It is

						propagated by seeds & cuttings
79.	*Clerodendrum variegatum*	-	Verbenaceae	-	-	-
80.	*Codiaeum variegatum*	Croton	Euphorbi-aceae	-	-	It is an evergreen bush with wide variety of size, shape & colour of leaves. Popular foliage plant grown in pot & also in ground
81.	*Coleus blumei*	-	Labiatae	-	-	A soft, perennial herb (or) sub-shrub growing to a height of 30cm – 1m. It requires adequate light. It can be grown in pot (or) in ground. It is very susceptible to wet soil (or) water stagnation even for a short period It is propagated by seeds & cuttings
82.	*Coleus rehneltianus*	-	Labiatae	-	-	-
83.	*Crossandra undulaefoliaSyn: C. infundibuliformis*	Fire craker plant	Acanthaceae	Orange yellow, deep orange, salmon pink	Throughout the year	It is a dwarf, evergreen shrub growing to a height of 1m .It is a quick growing shrub. It prefers warm, humid climate It is suitable for sun (or) semi-shade. It is a

						perpetual flowering shrub for shrubbery & in pot. Propagated by cuttings (or) seeds
84.	*Cuphea hyssopifolia*	False heather	Lythraceae	Purplish rose	-	A dwarf, bushy plant. It is a good pot plant. Propagated from cuttings
85.	*Cuphea melvillea*	-	Lythraceae	Red at base, yellow at open	-	A perennial sub shrub grows to a height of 1m.
86.	*Cuphea miniata*	-	Lythraceae	Bright red	-	It is a perennial, evergreen plant. It grows up to 30cm height. Propagated by cuttings. It is grown in shrubbery & also in pots.
87.	*Daedalacanthus macrophyllus*	-	Acanthaceae	Pale violet - blue	Winter & early summer	It grows to a height of 1.5m. It grows well in partial shade & full sun. Propagated by cuttings.
88.	*Daedalacanthus nervosus*	-	Acanthaceae	blue	Dec - Mar	It is a shrub growing to a height of 1m. It is an excellent plant for shrubbery & also in pot
89.	*Datura chlorantha*	-	Solanaceae	Yellow, fragrant	-	It is a tall glabrous shrub
90.	*Datura fastuosa*	-	Solanaceae	Violet outside, whitish within	-	A shrubby annual
91.	*Datura sanguine*	-	Solanaceae	Orange red	-	Flowers are of double type

92.	*Datura suaveolens*	Angel's trumpet	Solanaceae	White, fragrant	Summer & Rainy	It grows to a height of about 5m. It prefers full sun (or) partial shade. It is of quick growing. It is propagated by seeds & cuttings
93.	*Dombeya mastersii*	-	Sterculiaceae	Light rose	-	A much branched, coarse looking shrub growing to a height of about 2.5m
94.	*Dombeya natalensis*	-	Sterculiaceae	White, scented	-	A tall shrub with numerous spreading branches growing to a height of about 3-4m
95.	*Dombeya wallichi*	-	Sterculiaceae	Pink	Winter & early spring	A tall shrub growing up to 3m. It needs pruning to keep them in good shape. Light pruning should be done each year after flowering. Propagated from cuttings (or) air – layering
96.	*Duranta lorentzii*	-	Verbenaceae	White	-	A spineless, large shrub with drooping branches growing up to 3m height. It is an evergreen shrub. Propagated by cuttings
97.	*Duranta repens*	Golden dew drop	Verbenaceae	Blue	-	A tall, much branched, bushy shrub with axillary spine growing to a height of about 2.5 – 3m

98.	*Eranthemum alo-marginctum*	-	Acanthaceae	-	-	It is a bushy plant growing to a height of about 1.5m
99.	*Eranthemum bicolor*	-	Acanthaceae	Dark red	-	A well shaped medium, sized plant
100.	*Eranthemum laxiflorum*	-	Acanthaceae	Purple - rose	Year round flowering abundantly in rainy season	A dwarf flowering shrub growing to a height of 1m. It grows in partial shade
101.	*Eranthemum nigrum*	-	Acanthaceae	White with rosy centre	-	A well shaped bush about 1.5m
102.	*Eranthemum tricolor*	-	Acanthaceae	Rosy red	-	A popular foliage shrub grown in shade (or) semi-shade Colour of the foliage becomes dull in dense shade. Dry heat reduces leaf size & makes the colour less attractive. They are grown in pots & used for indoor decoration
103.	*Erythrina blakei*	-	Caesalpini-aceae	Deep red	Early summer to rainy season	A tall shrub growing to height of about 3m. It is a beautiful flowering shrub. Propagated from cuttings & air-layering
104.	*Erythirna crista-galli*	-	Caesalpini-aceae	Crimson - scarlet	Summer & rainy	A tall shrub (or) dwarf tree growing to a height of about

					4m with cracked whitish bark. It is propagated by cuttingsIt does not stand ill-drainage
105. *Erythrina resupinata*	-	Caesalpini-aceae	Bright scarlet	Feb - April	A deciduous low growing semi-woody plant. Propagated by seeds & cuttings
106. *Euphorbia cotinifolia*	Hierba mala	Euphorb-iaceae	Yellow	-	A tall, much branched shrub growing up to 3m. Frequent watering is necessary in summer months. It needs pruning for a good shape. Propagated from cuttings
107. *Euphorbia fulgens*	Scarlet plume	Euphorbi-aceae	Orange red	-	A small, handsome shrub, suitable in shrubbery, rock gardens & pots. Propagated from cuttings
108. *Euphorbia heterophylla*	Annual poinsettia	Euphorbi-aceae	-	-	It grows to a height of 1m. Leaves upper ones are small red at the base, bract like. It is hardy & quick growing plant for sunny location. Propagated from seeds & cuttings

109. *Euphorbia leucocephala*	White poinsettia	Euphorbi-aceae	Small yellow	-	Tall shrub with uniformly spreading thin, pinkish branches. Bracts are white in colour. It is a spectacular shrub when the branches are covered with snow-white bracts during winter months. It is grown in sunny locations. Pruning is done after the flowering is over. Propagated by cuttings.
110. *Euphorbia pulcherrima*	Chrismas flower	Euphorbi-aceae	-	Winter months	Tall (or) medium sized shrubs. Bracts colour vary from crimson, yellow, rose, white. It is grown in open, both in ground & pot. It can't withstand frost (or) very low temperature. It is a hardy, quick growing. Plants are pruned after flowering
111. *Evodia elegans*	-	Rutaceae	Small green	-	It is an evergreen shrub. It is grown for its attractive foliage & bushy growth growing to a height of about 2m. A well shaped plant for sun, semi-shade. Propagated by cuttings. It forms an attractive plants in pots.

112. *Excoecaria bicolor*	-	Euphorbiaceae	Yellow, small	-	A handsome bushy shrub growing to a height of about 1.5m. It thrives well in full sun & partial shade. It is propagated by cuttings. It does not respond to pruning
113. *Fortunella japonica*	-	Rutaceae	White, small, scented	Summer & rainy	A tall shrub growing to a height of 2m. It is a popular pot plant. Propagated by cutting (or) air-layering
114. *Gardenia jasminoides*	Cape Jasmine	Rubiaceae	White colour, gradually turning yellow, fragrant	Summer & early rainy	A popular well – shaped growing to a height of 3m. It prefers warm & humid regions
115. *Gardenia longistyla*	-	Rubiaceae	White	Hot months	An evergreen shrub with drooping branches. Propagated by cuttings
116. *Gardenia lucida*	-	Rubiaceae	White, scented	Summer & rainy	Large, bushy, evergreen shrub growing to a height of 3m. It can be grown as specimen plant (or) at the back of the shrubbery
117. *Gardenia radicans*	Miniature cardenia	Rubiaceae	White	Summer	A compact & busy shrub growing to a height of less than 1m. It is of slow growing. It is a dwarf shrub. It prefers full sun & semi-shade

					Suitable for growing in shrubbery, rock garden & in pots
118. *Gardenia thunbergia*	-	Rubiaceae	White, scented	Summer	It grows to a height of 2m. It grows well in full sun & partial shade. Propagated by cuttings & air-layering
119. *Gmelina asiatica*	-	Verbenaceae	Yellow	Summer & Rains	A tall, erect, spiny shrub. Plant is pruned after flowering. Propagated from cutting (or) air-layering
120. *Graptophyllum pictum*	Caricature plant	Acanthaceae	Deep red	-	A beautiful foliage shrub. It grows well in bright light but not in full sun, particular in dry summer. It is a handsome bush grown in ground & also in pot. It is propagated by cuttings.
121. *Gustavia insignis*	-	Lecythi daceae	Rose coloured & lower part creamy white	Year round	It is propagated by layering. It should be grown in partial sun. It is a slow growing
122. *Hamelia patens*	-	Rubiaceae	Orange scarlet	Year round	A handsome, perpetual flowering shrub growing up to 2.5m. It grows well both in sun & semi-shade. Propagated by cuttings.

123. *Hamiltonia suaveolens*	-	Rubiaceae	Pale heliotrope in colour	Winter	A deciduous shrub with light grey branches, growing to a height of 3m. It grows well in full sun & semi-shade. It is propagated by cuttings. Pruning is to be done after flowering
124. *Hibiscus colllinus*	-	Malvaceae	White with purplish brown centre	Summer & rainy	A tall & bushy shrub grows to a height of 4m. It is a large evergreen plant. Propagated by seeds, cuttings & air-layering
125. *Hibiscus mutabilis*	Changeable rose	Malvaceae	Petals white at first, changing to deep pink & finally red	Sep - Dec	A deciduous shrub grows to a height of 2.5m. It is multiplied by air-layering, cutting & seeds
126. *Hibiscus rosa-sinensis*	China rosse	Malvaceae	White, yellow, orange, red, pink, salmon & purple	Throughout the year	It is an evergreen shrub growing to a height of 1-3m. It is a good pot plant. It does not prefers water stagnation (or) softy soil. It prefers full sun & partial shade. Propagated by cuttings, layering, budding & grafting.
127. *Hibiscus schizopetalus*	Coral hibiscus	Malvaceae	Orange - red	Throughout the year	A tall, glabrous shrub (or) small tree growing to a

					height of 4m. It is very hardy. Propagated by cuttings.
128. *Hibiscus syriacus*	Rose of Sharon	Malvaceae	Rose, blue, purple, white, violet	Throughout the year	It is a deciduous, erect, much branched shrub growing to a height of 2m.
129. *Hydrangea* spp.	-	-	-	-	It is a medium tall growing shrub. It is grown as specimen plant (or) in the shrubbery border. It is an excellent pot plants for indoor & outdoor decorations. They prefer full sun (or) semi-shade. Propagated by cuttings. It responds to pruning.
130. *Hydrangea arborescens*	-	Saxifragaceae	Dull white	July-Aug	Plants grows up to 1.25m
131. *Hydrangea macrophylla*	-	Saxifragaceae	Pink (or) blue	-	It is known as common hydrangea. Flower is blue colour in acidic soil with aluminium & pink in soils of higher pH with calcium
132. *Hydrangea paniculata*	-	Saxifragaceae	Yellowish white	Summer to Autumn	A handsome tall plant grows up to 3m
133. *Iochroma tubulosum*	-	Solanaceae	Violet	Year round	A bushy shrub. It prefers full sunIt doesn't respond to

					pruning. Propagated by cuttings & air - layering
134. *Iresine herbstii*	Beefsteak plant	Amaranh-aceae	-	-	It is a bushy herb
135. *Iresine lindenii*	Blood leaf	Amaranth-aceae	-	-	A low spreading (or) semi-erect sub shrub. It is mostly grown as edge plants in the garden. It is propagated by cutting (or) seed
136. *Ixora acuminata*	-	Rubiaceae	White, fragrant	-	A tall glabrous shrub growing up to 3m high
137. *Ixora aliporensis*	-	Rubiaceae	Orange	-	It is a hybrid between *Ixora coccinea* &*Ixora venusta*
138. *Ixora barbata*	Bearded ixora	Rubiaceae	White	-	-
139. *Ixora chinensis*	Chinese ixora	Rubiaceae	Red	-	A dwarf species growing up to 15m high
140. *Ixora coccinea*	-	Rubiaceae	Deep scarlet	-	A compact shrub growing up to 1.5m tall
141. *Ixora duffii*	-	Rubiaceae	Orange scarlet	Summer & rainy	A highly floriferous species of *Ixora*
142. *Ixora fulgens*	-	Rubiaceae	Orange scarlet	-	It grows up to 2m high
143. *Ixora hybrida*	-	Rubiaceae	-	-	It has a cluster of buff-coloured flowers
144. *Ixora lutea*	-	Rubiaceae	Yellow	-	A straggling shrub with slender branches

145. *Ixora macrothyrsa*	-	Rubiaceae	Bright scarlet	-	It is a large, tall plant with woody grows up to 3m
146. *Ixora profuse*	—	Rubiaceae	Yellow	-	An erect, compact, bushy shrub of about 2m high
147. *Ixora spectabilis*	-	Rubiaceae	Bright orange	-	It is a shrub growing up to 1.5m high
148. *Ixora superba*	-	Rubiaceae	White, scented	-	A dwarf, very slow growing shrub with few stout woody branches
149. *Ixora venusta*	-	Rubiaceae	Pale pink	-	Propagated by cuttings. It is a tall shrub growing up to 2m high. It prefers full sun. It grows as pot plant in shrubbery (or) as specimen plant.
150. *Jacobinia carnea*	Flamingo plant	Acanthaceae	Pink	Throughout the year	An erect shrub, up to 2m high. Propagated from cuttings. It responds to pruning
151. *Jacobinia gheisbreghtiana*	-	Acanthaceae	Scarlet	Summer	It is a semi-woody, bushy growing to a height of about 1.5m. It should be grown in partial shade
152. *Jacobinia pauciflora*	-	Acanthaceae	Scarlet with yellow tips	Throughout the year	It is a dwarf shrub growing to a height of about 60cm. It

					prefers full sun (or) seni-shade
153. *Jacquinia ruscifolia*	-	Theophras-taceae	Orange turning reddish brown	Summer	It is a handsome evergreen, tall shrub (or) small tree growing to a height of 4m. It is a well shaped, slow growing plant. Propagated from seeds
154. *Jasminum humile*	Yelloe jasmine	Oleaceae	Yellow, fragrant	Winter, summer & spring	A semi erect shrub. Propagated by cuttings. It is an evergreen, scandent shrub
155. *Jasminum pubescens*	-	Oleaceae	-	-	-
156. *Jasminum sambac*	Arabian jasmine	Oleaceae	White, fragrant	Summer & rainy	A dwarf, compact bushy shrub growing to a height of 1m with pubescent branches. They are widely grown in pots
157. *Jatropha panduraefolia*	-	Euphorb-iaceae	Red	-	A perpetual flowering shrub growing to height of 3m. It is a quick growing shrub. It is suitable for full sun (or) partial shade. It responds to light pruning. It is propagated by cuttings (or) air-layering

158. *Justicia aurea*	-	Acanthaceae	Yellow	Summer	A semi woody erect bush growing up to 1.5m. Propagated by cuttings.
159. *Justicia gendarussa*	-	Acanthaceae	Purple spots	summer	It is a popular hedge plant. Quick growing shrub. It is propagated by cuttings.
160. *Justicia ovata*	-	Acanthaceae	Orange - scarlet	Many months except in winter	A free flowering dwarf shrub growing up to 80cm. It is a free flowering dwarf shrub. Suitable for sun & semi-shade Propagated by cuttings.
161. *Kopsia fruticosa*	-	Apocynaceae	Reddish white & tip and centre of the flower is red	Throughout the year	It has a uniformly spreading habit. Propagated by cuttings of terminal leaf shoot.
162. *Lagerstroemia indica*	Crape myrtle	Lythraceae	White, rose, pink, mauve	May - August	A much branched woody, erect, large shrub growing.
163. *Lagerstroemia lancasteri*	-	Lythraceae	Mauve	May - Aug	It is a large, woody shrub growing up to 3m. Pruning should be done during the month of Feb. It is propagated by cuttings.
164. *Lantana camara*	-	Verbenaceae	Colour of flower, often change with age	-	Prickly – stemmed, bushy planting of 60cm – 2m height.

165. *Lantana sellowiana*	-	Verbenaceae	Mauve on flat heads	Cold months	It can't stand water stagnation. Used in hanging basketsA trailing shrub. It is an ideal plant for rock garden. It is grown shrubbery, rock gardens, as ground cover & hedge.
166. *Lawsonia inermis*	Tree mignonette	Lythraceae	-	Summer & rainy	A tall shrub growing up to 4m. It is also grown as tall hedge It is pruned after flowering. Propagated from seeds (or) cuttings
167. *Leea alata*	-	Vitaceae	Bright crimson	Rainy	A large semi-woody shrub growing up to 2m high. Propagated by seeds, suckers (or) cuttings. It flowers well in semi-shade.
168. *Ligustrum robustrum*	-	Oleaceae	White	-	A large shrub with spreading branches growing up to 3m. It comes up well in shady location. Propagated by seeds & cuttings.
169. *Lindenia rivalis*	-	Rubiaceae	White, scented	Summer	It is tall shrub growing to a height of about 2.5 – 3m high
170. *Magnolia fusctata*	-	Magnoliaceae	Yellowish green, stained brown purple	Summer & rainy	It is a bushy shrub growing to a height of about 2m. It is suitable to grow in partial

					shade. Propagated by cuttings & air-layering
171. *Magnolia mutabilis*	-	Magnoliaceae	Yellow, highly scented	Summer &rainy	It grows up to 2m high. It prefers semi-shade. Propagated by green wood cuttings
172. *Magnolia pumila*	-	Magnoliaceae	Creamy white	-	An evergreen bushy plant, grows up to 2m high. It is a slow growing plant. Propagated by cuttings. It prefers semi-shade
173. *Malpighia coccigera*	-	Malpighi-aceae	-	-	-
174. *Malpighia glabra*	Barbados cherry	Malpighi-aceae	-	-	-
175. *Malvaviscus arboreus*	-	Malvaceae	Scarlet	Year round	It grows up to 2.5m. It is a popular shrub in tropical gardens. It is grown in the shrubbery as hedge & also in pots. It prefers full sun & shadeIt responds to pruning. Propagated by semi-woody cuttings
176. *Manihot esculenta*	-	Euphorbi-aceae	-	-	A quick growing attractive shrub. It prefers full sun. It grows well in ground & pot. Propagated by cuttings.

177. *Melastoma decemfidum*	-	Melastom-aceae	-	-	-
178. *Melastoma malabathricum*	-	Melastom-aceae	Mauve-purple	Winter & spring	It is a hardy shrub for shrubbery & rock garden. Propagated by cuttings
179. *Melastoma sanguineum*	-	Melastom-aceae	-	-	-
180. *Memecylon edule*	Red iron wood	Melastom-aceae	Blue	-	It is a tall, well, shaped shrub growing to a height of 3m.
181. *Memecylon ellipticum*	-	Melastom-aceae	Pinkish blue	Mar - May	It is a spreading well shaped woody shrub growing up to 2.5m high. It is a slow growing, woody, evergreen shrub. Propagated from seeds & air-layering
182. *Muehlenbeckia platyclada*	-	Polygonaceae	Greenish white	Summer & rainy	An evergreen, glabrous shrub of attractive appearance growing to a height of 1.5cm high. It grows well in full sun & partial shade. It is grown in pots for indoor decoration. Propagated by cuttings
183. *Murraya exotica*	Chinese box	Rutaceae	-	Summer & rainy	A tall, evergreen shrub with a grayish ark. It is grown in

					shrubbery, as a specimen plant. It thrives well in sun & semi-shade. It is grown as hedge. It does not need regular pruning. Propagated by seeds, cuttings & air - layering
184. *Murraya paniculata*	-	Rutaceae	-	-	-
185. *Mussaenda erythrophylla*	-	Rubiaceae	Bright scarlet above whitish with red veins & below hairy on both surfaces	Summer & rainy	A semi erect shrub growing to a height of 2m. It prefers semi-shade. Propagated by cutting (or) layering
186. *Mussaenda frondosa*	-	Rubiaceae	Orange, expanded single sepal, creamy white	Summer & rainy	An ornamental shrub with spreading habit. Propagated by cutting & air - layering
187. *Mussaenda luteola*	-	Rubiaceae	Yellow	April - Sep	An erect, slender shrub growing up to 2.5m high. It is propagated by cutting
188. *Mussaenda philippica 'Aurorae'*	-	Rubiaceae	Yellow	April - Oct	A tall deciduous shrub growing to a height of about 2.5m. A large woody shrub with spreading branches. It prefers full sun & partial shade. It is grown shrubbery, as specimen plant & also in pots

189. *Nandina domestica*	Sacred Chinese bamboo	Berberidaceae	White	-	It is a handsome evergreen shrub growing to a height of 2m. It prefers semi-shade. It is suitable to grow in pot & also in ground. It is propagated by seed, cutting & division of clumps
190. *Nerium oleander*	Oleander	Apocynaceae	Pink, white, carmine	Summer & rainy	It is an evergreen erect shrub growing to a height of about 2.5m. Propagated by cuttings It is a tall flowering shrub
191. *Ochna kirkii*	-	Ochnaceae	Yellow	Year round, more profusely in summer	It is an evergreen plant growing to height of 2.5m. It is a slow growing shrub
192. *Ochna squarrosa*	-	Ochnaceae	Yellow	-	A tall, woody, deciduous shrub growing to a height of about 3m. Propagated from seeds & cuttings
193. *Oncoba spinosa*	-	Flacourtia-ceae	White	Summer	A large woody shrub (or) small tree growing up to 4m high. Propagated from seeds, cuttings & layering
194. *Osmanthus fragrans*	-	Oleaceae	White, strongly fragrant	Throughout the year, profusely during summer	A slow growing shrub. It prefers semi-shade. Propagated by cutting & layering

195. *Pachystachys lutea*	Lollipop plant	Acanthaceae	White, emerging through golden yellow, showy bract	-	A branched, semi-woody plant growing to a height of 50cm. A well shaped plant. Propagated by cuttings. It prefers full sun & semi-shade
196. *Pentas lanceolata*	Egyptian star cluster	Rubiaceae	Star cluster	-	A semi-woody, perpetual flowering shrub growing to a height of 40-60cm in height. It grows as pot plants also. It prefers semi-shade (or) bright light. Propagated by cuttings. It responds to light pruning
197. *Pentas karmesiana*	-	Rubiaceae	-	-	A woody erect shrub growing to a height of 1.5m height. Propagated by cuttings. It shows bushy appearance
198. *Petrea arborea*	-	Verbenaceae	Purplish blue	Summer & rainy	It is of slow growing. It thrives well in full sunny location. Propagated by tip cuttings.
199. *Phyllanthus nivosus*	Snow bush	Euphorbiaceae	Greenish	-	A busy shrub growing up to 2m high. Semi-deciduous plant. It prefers full sun. Propagated by seeds (or) seed cutting. It is grown as popular pot plant.

200. *Plumbago auriculata*	Cape leadwort	Plumbagi-naceae	Azure blue	10 months in a year	A dwarf, bushy sub-scandent shrub growing up to 1m height. It is grown in shrubbery, rock garden & pots. It is very susceptible to water logging. It is propagated by cuttings. It can be pruned in rainy season
201. *Plumbago indica coccinea*	Scarlet leadwort	Plumbag-inaceae	Red	Winter	It is a dwarf shrub growing to a height of 50cm. It grows well in full sun & partial shade. It is grown as a pot plant. Propagated by cutting. Pruning is done during rainy season
202. *Pogonopus exsertus*	-	Rubiaceae	Pink	Throughout the year	It is a shrub with terete branches. Propagated by cuttings & layering. Pruning should be done after flowering.
203. *Polyscia balfouriana*	-	Araliaceae	-	-	It is a compact & bushy shrub growing to a height of about 2m
204. *Polyscia filicifolia*	-	Araliaceae	-	-	It is an erect, glabrous shrub growing to a height of about 2m (or) more

205. *Polyscia fruticosa*	-	Araliaceae	-	-	It is an evergreen shrub growing to a height of about 2m
206. *Polyscia guilfoylei*	-	Araliaceae	-	-	An erect bushy, glabrous shrub growing up to 3m height. It is much branched
207. *Polyscia scutellaria*	-	Araliaceae	-	-	Large shrub, leaves cordali (or) oblong, simple. A large specimen plant grown in pot. It is suitable for growing in semi shade in ground & light It can grow well in shade. Propagated by tip cuttings. Pruning in early rains will help to maintain the shape of plant
208. *Pootia grandiflora*	-	Apocynaceae	White (or) creamy, scented	Summer & rainy	A tall, spreading shrub with long branches growing to a height of about 4m. It is recommended for large shrubbery. Propagated by cuttings (or) air - layering
209. *Portlandia grandiflora*	-	Rubiaceae	Large white, pinkish inside the throat	Summer & rainy	It is a beautiful, tall, glabrous shrub growing to a height of about 2.5m. It produces funnel shaped, large flowers. It is grown in shrubbery & as

					specimen plant. Propagated by cuttings
210. *Posoqueria latifolia*	Needle flower	Rubiaceae	-	Mar - May	A large evergreen shrub reaching a height of about 4m. It is a tall & spreading shrub. It is suitable to plant at the back of a large shrubbery. It responds to pruning very well. Propagated by cuttings & air-layering
211. *Pseuderanthemum alatum*	Chocolate plant	Acanthaceae	-	-	It is a handsome low growing herb. It is grown in pot. Propagated from cuttings
212. *Pseuderanthemum atropurpureum*	-	Acanthaceae	Purpling with red marking, small size	-	A glabrous shrub growing to a height of about 1.2m high
213. *Pseuderanthemum reticulatum*	-	Acanthaceae	-	-	It grows to a height of 1m with angled glabrous brachlets. It is a hardy shrub. It grows very well in sun & semi-shade
214. *Pseuderanthemum sinuatum*	-	Acanthaceae	White, spotted rosy	-	A bushy, semi-woody shrub growing up to 60cm high
215. *Punica granatum*	Pomegranate	Punicaceae	-	-	-

216. *Quassia amara*	-	Simarubaceae	Scarlet	Mar - July	A tall shrub growing up to 2.5m high. It grows well in semi-shady. Propagated by cuttings
217. *Randia macrantha*	-	Rubiaceae	White, large turning pale yellow	Summer	A well shaped shrub with uniformly spreading branches. It is a slow growing much branched shrub growing up to 2.5m
218. *Randia maculate*	-	Rubiaceae	-	-	-
219. *Ravenia spectabilis*	-	Rutaceae	Crimson pink	-	An evergreen woody shrub of 2.5m high with spreading branches. It grows well in sun & semi-shade. Propagated by air-layering (or) cutting
220. *Reinwardtia trigyna*	-	Linaceae	Yellow	Year round	A sub shrub, semi-woody growing to a height of 50cm It is a showy, dwarf bush. It is grown in shrubbery & also in pot. Propagated by cuttings
221. *Rondeletia odorata*	-	Rubiaceae	Vermillion with conspicuous, yellow throat	7-8 months in a year	A medium sized, woody shrub
222. *Rondeletia strigosa*	-	Rubiaceae	Bronzy crimson with orange throat	7-8 months in a year	A small, well shaped bush with many small woody branches. It thrives well in

					full sun & partial shade. Propagated by cutting (or) air-layering
223. *Ruellia lorenziana*	-	Acanthaceae	Bluish purple	Summer	A small glandular herb growing up to 40cm height. Stem is quadrangular. A quick growing herb. It grows very well in full sun & semi-shade. Propagated from cuttings & seeds
224. *Ruellia rosea*	-	Acanthaceae	Bright red	Throughout the year	It is a low growing herbaceous perennial. It prefers sun (or) semi-shade. Propagated by cuttings.
225. *Russelia juncea*	Coral plant	Scrophula-riaceae	Bright red	Summer & rainy	A much branched plant of unusual appearance, almost prostrate with few erect branches. It is a grown in shrubbery, rock garden & pots. Propagated from cuttings & division of clumps
226. *Russelia sarmentosa*	Coral blow	Scrophula-riaceae	Red	-	It is an erect bushy shrub growing to a height of about 60-70cm. It grows very well in full sun. Propagated from cuttings

227.	*Ruttya fruticosa*	-	Caesalpini-aceae	Orange red with a black blotch at the centre	Throughout the year	A small shrub with woody branches. It can be grown in sun & semi-shade. Propagated from cuttings.
228.	*Sambucus canadensis*	American elder	Sambucaceae	White, small	Summer & early rains	It is a tall shrub growing up to 4m. Propagation from seeds, hardwood cuttings & suckers
229.	*Sanchezia nobilis*	-	Acanthaceae	Yellow, large	Summer & rainy	A quick growing plant with showy foliage & long spikes of flowers. It is grown in pot as foliage plant. It prefers semi-shade. Propagated by cuttings
230.	*Sarcocephalus cordatus*	-	Rubiaceae	White	April - May	A large shrub growing up to 3m height. It is an evergreen plant. It produces branches from lower part of the stem. Propagated by seeds & cuttings
231.	*Serissa foetida*	-	Rubiaceae	White, light pink beneath	Summer & rainy	A dwarf & bushy shrub growing up to 1.5cm height. It is grown in shrubbery, rock gardens. Propagated from cuttings
232.	*Solanum macranthum*	Brazilian potato tree	Solanaceae	Bluish violet fading to white	Throughout the year	A tall shrub with spreading branches growing up to 4m

						height. Yellowish prickles are found on stem & petiole. It isa quick growing large shrub. It is grown as specimen plant. It responds to pruning very well
233.	*Solanum rantonnetti*	-	Solanaceae	Violet with yellow centres	April - Aug	It is an erect shrub growing up to 1.5m height. Propagated by cuttings
234.	*Stachytarpheta indica*	-	Verbenaceae	Blue on terminal slender spike	Summer & rainy	An evergreen, dwarf, bushy shrub. Branches are almost quadrangular
235.	*Stachytarpheta mutabilis*	-	Verbenaceae	Crimson, fading to rose	Summer & rainy	A semi-woody perennial, bushy plant growing up to 1.5m high. Branches are quadrangular. It is a hardy, quick growing shrub. Propagated by cuttings & seeds
236.	*Stemmadenia bella*	-	Apocynaceae	White, large, s cented yellow inside	Summer & early rains	It is a large, evergreen shrub with spreading branches growing up to 4m. It appears like a small tree. It grows very well in sun & semi- shade. Propagated by cuttings
237.	*Streptosolen jamesonii*	-	Solanaceae	Yellow turning reddish orange,	Throughout the year	It is an evergreen woody scabrous shrub growing up

			funnel shaped		to 2m high. It is not widely grown in gardens. Propagated by green, semi-woody cuttings
238. *Strobilanthes dyerianus*	-	Acanthaceae	Violet – blue, funnel shaped	-	A colourful foliage shrub growing up to 1m high. It thrives well in sun & semi-shady conditions. Propagated by cuttings
239. *Strobilanthes glomeratus*	-	Acanthaceae	Bluish purple, funnel shaped	Summer & rainy	A dwarf shrub with horizontal, quadrangular branches growing up to 1.5m height. Showy bracts look like flowers. Propagated by cuttings
240. *Strophanthus dichotomous*	-	Apocynaceae	White	Summer & rainy	A large shrub with semi-scandent branches
241. *Strophanthus wallichii*	-	Apocynaceae	Yellow	Summer & rainy	A medium sized shrub with slender woody branches
242. *Tabernaemontana coronaria*	Cape jasmine	Apocynaceae	White	Year round except in cold months	A very popular large shrub in gardens. Propagated by cutting. It is grown as hedge & also for screening Unpalatable to cattle. It responds to pruning

243. *Tabernaemontana dichotoma*	Eve's apple	Apocynaceae	White scented	Hot months	Tall, bushy, evergreen plant
244. *Tarenna zeylancia*	-	Rubiaceae	White, highly scented	Throughout the year, profuse flowering in summer	It is a tall woody shrub. It is planted in shrubbery. Propagated by cuttings.
245. *Tecoma gaudichaudi*	-	Bignoniaceae	Golden yellow	Throughout the year	A large, erect shrub grows up to 4m height. A quick growing tall shrub. It is very susceptible to water logging. It responds to pruning. Propagated by cuttings (or) seeds
246. *Tecoma stans*	Yellow bell	Bignoniaceae	Yellow	Summer & rainy	An erect shrub with spreading branches grows up to 3m. It is a tall, hardy shrub. It prefers sun & semi-shade. Propagated by seeds & cuttings
247. *Tecomaria capensis*	Cape-honeysuckle	Bignoniaceae	Orange (or) scarlet	Throughout the year	It is a tall, bushy shrub with woody & spreading branches It responds to pruning Propagated from cuttings
248. *Thryallis glauca*	-	Malpighi-aceae	Yellow profusely during summer months	Year round	It is a bushy, quick growing shrub grows up to 2m high. It responds to pruning.

					Propagated from seeds & cuttings.
249. *Thunbergia erecta*	-	Acanthaceae	Dark violet, orange throat	Throughout the year profusely during hot months	It is a hardy, erect & bushy shrub. It is deciduous during winter. It thrives well in sun & semi-shade. Propagated from soft & hardwood cuttings
250. *Thunbergia hybrida*	-	Acanthaceae	Blue-violet	Summer & rainy	A straggling bush with many semi-woody branches. It flowers very well in full sun & semi-shade. Propagated by cuttings
251. *Thunbergia natalensis*	-	Acanthaceae	Pale blue	Summer & rainy	It is a dwarf bushy shrub growing up to 60cm height. It is a hardy shrub. Propagated by hardwood cuttings
252. *Turnera ulmifolia*	Sage rose	Turneraceae	Sulphur yellow	Throughout the year & more abundant during summer & rainy	A dwarf, semi-erect shrub up to 60cm high. Flowers are funnel shaped. It is suitable for growing in shrubbery & rock garden. It grows very well both in sun & semi-shade situations. Propagated by cuttings
253. *Vinca major*	Madagascar periwinkle	Apocynaceae	-	-	-

254. *Vinca minor*	-	Apocynaceae	-	-	-
255. *Vinca rosea*	-	Apocynaceae	Rose (or) white	Year round flower	A dwarf, much branched bushy shrub growing up to 60-80cm high. It is a hardy shrub. It is grown in pots, rock harden & in shrubbery. It prefers full sun
256. *Vitex agnus-castusSyn: V. negundo*	Chaste tree	Verbenaceae	Mauve blue (or) silver blue	Mild summer until autumn	A hardy, deciduous shrub of spreading habit growing up to 3m. In winter the leaves shed off
257. *Wedelia trilobata*	-	Compositate	Yellow	Throughout the year	A dwarf shrub growing to a height of about 60cm. It is grown as an under-shrub in shrubbery & rock gardens. Propagated by cuttings
258. *Woodfordia floribunda*	-	Lythraceae	Reddish orange	Feb - April	A much branched woody shrub growing up to 2.5m highPropagated by cuttings.
259. *Wormia burbidgii*	-	Dilleniaceae	Yellow	Summer & rainy	A large, spreading shrub growing up to 2.5m high. Leaves are large & shiny. It prefers partial shade. Propagated by cuttings & air-layering

Chapter **6**

Climbers & Creepers

I. General information

1. Climbers which climbs up with the help of **tendrils** are ***Antigonon leptopus, Bignonia gracilis, Pyrostegia venusta, Clematis paniculata***
2. Climbers which climbs up with the help of **thorns or prickles** are ***Bougainvillea* spp., climbing rose**
3. Climbers which climbs up with the help of **root and rootlets** are ***Campsis grandiflora, Ficus repens***
4. A climber which climbs up with the help of **sticky substances** from growing points is ***Ficus repens***
5. A Climber which climbs up with the help of **hookers** is ***Combretum comosum***
6. Climbers which climbs up with the help of **clingers** are ***Passiflora caerulea, Passiflora quadrangularis* and *Hydrangea petiolaris***
7. Climbers which grow well under **sunny situation** are *Pyrostegia venusta, Quisqualis indica, Antigonon leptopus, Adenocalymma alliaceum, Bougainvillea spp., Campsis grandiflora* and *Wisteria sinensis*
8. Climbers which grow well under **partial shade** are *Aristolochia* spp., *Beaumontia grandiflora, Clerodendrum splendens, Echites caryophyllata, Hiptage madablota, Hoya carnosa, Petrea volubilis, Lonicera japonica, Trachelospermum jasminoides, Quisqualis indica, Solanum jasminoides, S. seaforthianum, S. wendlandii, Strophanthus gratus*
9. Climbers which are grown for **screening purpose** are *Antigonon, Clerodendrum splendens, Derris scandens, Echites caryophyllata, Ipomea spp., Melodinus monogynus, Stictocardia tilifolia, Thunbergia laurifolia, T. mysorensis, Vallaris heynei, Bougainvillea* spp., *Pyrostegia venusta* and *Vernonia elaeagnifolia*

10. Climbers which are grown for **indoor decoration** are *Epipremnum aureum, Monstera deliciosa, Philodendron spp., Asparagus spp.*
11. Climbers which is grown in **pots** are *Clematis flammula, Clerodendrum thomsonae, Hoya carnosa, H. bella, Bougainvillea* spp., *Jacquemontia violacea, Solanum seaforthianum, Tristellateia australis, Stephanotis floribunda, Thunbergia alata, Quamoclit pinnata, Ipomea* spp.and *Clitoria ternata*
12. Climbers which are grown in **porches** are *Clerodendrum splendens, Ipomea horsfalliae, Pyrostegia venusta, Petrea volubilis, Jasminum* spp., *Bougainvillea* Mary Palmer, Thimma, Partha, De, R. R. Pal, Formosa and Refulgens
13. Climbers which are grown in **pergolas** are *Allamanda cathartica, Camoensia maxima, Chonemorpha macrophylla, Congea tomentosa, Derris scandens, Delima sarmentosa, Quisqualis indica, Thunbergia mysorensis, Pseudocalymma alliaceum, Rhynchodia wallichii* and *Wisteria sinensis*
14. Climbers which are grown **for low wall and trellis** are *Lonicera* spp., *Gloriosa, Clytostoma purpureum, Clematis flammula, Campsis grandiflora, C. radicans, Cobaea scandens, Clerodendrum thomsonae, Pandorea jasminoides, Solanum jasminoides, Stephanotis floribunda, Thunbergia alata, Quamoclit pinnata, Quamoclit lobata* and *Lathyrus odoratus*
15. Climbers which are grown **for hedging** are *Bougainvillea* spp. and *Clerodendron inerme*
16. Climbers which produce **fragrant flowers** are *Jasminum grandiflorum, Jasminum officinale, Trachelospermum jasminoides, Solanum jasminoides, Tecoma jasminoides, Clematis paniculata* and *Hiptage benghalensis*
17. Heavy climbers are the climbers which **produce luxuriant vegetative growth** and grow very vigorously. They cover large area and suitable for covering bigger area
18. Examples for **heavy climbers** are *Antigonon leptopus, Bauhinia vahlii, Beaumontia grandiflora, Bignonia magnifica, Bougainvillea spp., Hiptage benghalensis, Quisqualis indica, Pyrostegia venusta, Clerodendrum splendens, Combretum comosum, Cryptostegia grandiflora, Derris scandens, Pseudocalymma alliaceum, Thunbergia grandiflora, Vallris heynei* and *Wisteria sinensis*
19. Light climbers are the climbers which make sparse vegetative growth and remain light in spread
20. Examples for light climbers are *Bignonia unguiscati, Ipomea* spp., *Lonicera japonica, Jacquemontia violacea, Tristellateia australis, Stephanotis floribunda, Clerodendrum thomsonae, Quamoclit pinnata, Q. lobata, Tecoma jasminoides,* and *Clitoria ternatea*

21. Medium climbers are not fast growing like heavy climbers and the growth is much less than the light climbers but neatly cover a fairly large support, if trained properly

22. Examples for **medium climbers** are *Aristolochia elegans, A. ringens, Adenocalymma comosum, Banisteria laevifolia, Campsis radicans, Clematis gouriana, Clytostoma purpureum, Ipomea horsfalliae, Pandorea jasminnoides, Pyrostegia venusta, Podranea ricasoliana, Stigmaphyllon ciliatum, Trachelospermum jasminoides,*

23. Examples for **annual climbers** are *Cardiospermum halicacabum, Lathyrus odoratus,* nasturtium, *Ipomea lobata, Ipomea rubro-caerulea, I. multifida, I. alata, Cobaea scandens, Asarina erubescens, Thunbergia alata, Quamoclit pinnata, Quamoclit lobata* and *Clitoria ternatea*

24. Examples for **perennial climbers** are *Akebia quinata, Celastrus scandens, C. orbiculatus, Parthenocissus tricuspidata, Clematis* spp., *Aristolochia macrophylla, Hedera helix, Vitis coignetiae, Passiflora* spp., *Ampelopsis brevipedunculata, Rosa* sp., *Polygonum aubertii, Campsis radicans, Euonymous fortunei, Wisteria* spp.

25. Climbers which are planted **for covering slopes** are *Lonicera japonica, Thunbergia grandiflora* and *Passiflora caerulea*

26. Climbers which are planted for **making topiary** are *Clerodendrum inerme* and *Bignonia unguiscati*

27. Climbers which produce **ornamental fruits** are *Dioscorea deltoidea, Hedera nepalensis* and *Rosa* spp.

28. **Foliage climbers** are *Scindapsus aureus, S. pictus, Philodendron* spp., *Syngonium podophyllum, Monstera deliciosa, Cissus discolor, Asparagus* spp.

29. Planting distance for *Vernonia elaeagnifolia* is 15 - 20 cm

30. Planting distance for *Ficus pumila* is 8 - 10 cm

31. Common name for climbers are as follows

Allamanda violacea	:	Purple allamanda
Antigonon leptopus	:	Coral vine, Chain of love
Aristolochia elegans	:	Calico flower
Aristolochia ornithocephala	:	Bird's head flower
Beaumontia grandiflora	:	Nepal trumpet creeper
Clerodendrum thomsonae	:	Bleeding heart
Clitoria ternatea	:	Mussel- shell climber, butterfly pea creeper, Sangu pushpam

Cobaea scandens	:	Cup and saucer vine, purple-bell climber
Cryptostegia grandiflora	:	Indian rubber vine
Clematis paniculata	:	Virgin flower
Campsis grandiflora	:	Trumpet climber
Ficus pumila	:	Creeping fig, Indian Ivy
Ficus radicans	:	Rooting fig
Hoya bella	:	Miniature wax plant
Hoya carnosa	:	Wax plant
Ipomea learii	:	Blue dawn flower
Ipomea purpurea	:	Common morning glory
Ipomea palmata	:	Railway creeper
Ipomea rubro-caerulea	:	morning glory
I. alata	:	Moonflower vine
Jasminum grandiflorum	:	Spanish jasmine, Chameli, Jathimalli, Pitchi
Lathyrus odoratus	:	Sweet pea
Lonicera japonica	:	Japanese honeysuckle
Lonicera sempervirens	:	Trumpet honeysuckle
Passiflora quadrangularis	:	Giant granadilla
Passiflora coccinea	:	Red passion flower
Passiflora laurifolia	:	Watch flower, Phoolghari
Pyrostegia venusta	:	Golden shower
Petrea volubilis	:	Purple wreath
Porana paniculata	:	Bridal creeper, Bridal bouquet
Quamoclit coccinea	:	Star ipomea
Quisqualis indica	:	Rangoon creeper, Jhomka Bel
Solanum jasminoides	:	Potato vine
Solanum seaforthianum	:	Potato creeper
Solanum wendlandii	:	Giant potato creeper
Stephanotis floribunda	:	Madagascar jasmine, Creeping tuberose
Strophanthus gratus	:	Cream fruit
Thunbergia alata	:	Black-eyed susan
Thunbergia grandiflora	:	Sky flower
Trachelospermum jasminoides	:	Star jasmine
Wisteria chinensis	:	Chinese wisteria

32. Climbing plant which belongs to the family Annonaceae is

S. No.	Botanical Name	Common name	Flower colour	Flowering season	Remarks
1.	*Artabotrys odoratissimus*	Tail grape/Climbing ylangylang	Greenish yellow, strong fragrance	Hot & rainy season	It is an evergreen, large, shrubbery climber. Propagatedby seeds (or) layers

33. Climbing plant which belongs to the family Apocynaceae are

S. No.	Botanical Name	Common name	Flower colour	Flowering season	Remarks
1.	*Allamanda cathartica*	Yellow allamanda	Yellow, large	Summer & Rains	It prefer sunny location. It can be trained on wall, porch, pillar & arch. Propagated by cuttings
2.	*Allamanda violacea*	Purple allamanda	Light violet, large	Summer & rainy	A scrambling shrub with terete branches. It prefers a sunny location. It can be trained on low wall (or) pillar. Propagated by cuttings
3.	*Beaumontia grandiflora*	Nepal trumpet climper	Large, white	Jan - March	It is a heavy, evergreen climber. It is a quick growing. It flowers well in both sun & semi-shade. Propagated by cutting & layering

4.	*Chonemorpha macrophylla*	Frangipani vine	White yellowish in the throat, fragrant	Summer & rainy	An evergreen, heavy climber. It is suitable to grow on pergola or other strong support. Propagated by cutting & layering
5.	*Echites caryophyllata*	Malati	White, scented	Late summer & rainy	A heavy climber. It grows well in semi-shade & full sun. Propagated by cutting & layering Ideal climber on pergola, arch, wall & arbour in a sunny location
6.	*Echites rubro-venosa*	-	-	-	A slender climber. It grows well in partial sunlight, warm & moist conditions. It is a good plant for growing on pots with suitable support.
7.	*Melodinus monogynus*	East Indian Melodinus	White, scented	March - May	It grows well on strong arch & pergola It prefers full sun Propagated by cutting
8.	*Odontadenia grandiflora*	Odontadenia	Bright yellow, large, scented	Throughout the year	It is a large, shrubby climber. It prefers full sun & semi-shade. It is often grown on wall & pergola. Propagated from cutting & layering
9.	*Rhynchodia wallichii*	-	Yellowish white with a reddish centre	Summer & rains	A large, woody, evergreen climber. It grows well on pergola & wall. It serve for

					the purpose of screening. Propagated by cutting.
10.	*Strophanthus grandiflorus*	-	Light yellow, streaked with red, funnel shaped	Summer months	Large, woody climber. It is an extensive evergreen climber. Suitable for large gardens.
11.	*Strophanthus gratus*		White, tinted purple, scented	Summer & rainy	A large evergreen woody climber. It requires strong support & enough space. It is suitable on wall, pergola. Propagated by cuttings.
12.	*Trachelospermum jasminoides*	Chinese star jasmine, Confederate jasmine	White small, fragrant	Hot months	A large evergreen climbing shrub with white latex. It prefers full sun & partial shade. It is suitable to grow on fence, wall, trellis. Propagated by cutting (or) layering.

34. Climbing plant which belongs to the family Aristolochiaceae are

S. No.	Botanical Name	Common name	Flower colour	Flowering season	Remarks
1.	*Aristolochia elegans*	Calico flower	Corolla tube is yellow – green, flower is purple & white – blotched, white on the exterior, yellow eye	Hot months	Curious pouch flower. It is an ideal light climber. Propagated from cuttings.

2.	*Aristolochia grandiflora*	Pelican flower/swan flower	Large flowers with offensive odour, purple – blotched & veined	Oct - Nov	A shrubby, extensive climber. It is a heavy climber, needs strong support. It should be grown away from house. Propagated by seeds, cutting & layering
3.	*Aristolochia ornithocephala*	The bird's head flower	Netted (or) spotted dark purple, tube purple, upper lip violet inside, pouch-like, often shining dark purple	-	It is a free flowering climber. It is grown on walls & other strong supports. It prefers sunny locations. Propagated from cuttings of matured wood (or) by seed.
4.	*Aristolochia ringens*	Gaping Dutchman's Pipe	Pale green in colour, hairy within	Mar – May & also in Oct - Dec	It is a large evergreen climber. It is a quick growing climber. It is suitable for growing on wall
5.	*Aristolochia tomentosa*	Wooly Dutchman's Pipe	Yellow (or) yellowish green	-	A tall climber. It is quite har dy, grows to a great height. It is useful for the purpose of screening

35. Climbing plant which belongs to the family Asclepiadaceae are

S. No.	Botanical Name	Common name	Flower colour	Flowering season	Remarks
1.	*Hoya bella*	Miniature wax plant	Waxy white with purple centre	Summer	A slender, bushy plant with drooping branches. A small

					attractive plant. It can be grown as shrub (or) climber with light support It is grown in pots
2.	*Hoya carnosa*	Wax plant	Pinkish white with a red, star shaped crown, fragrant	Hot months	A vigorous growing twines attached to support by means of roots. It prefers partial shadeIt grows well in pot on suitable support. Propagated by cutting
3.	*Hoya imperialis*	Honey plant	Large, waxy dull - pink	Hot months	A tall, spreading climber. It produces cluster of beautiful flowers. Propagated by cuttings

36. Climbing plant which belongs to the family Bignoniaceae are

S. No.	Botanical Name	Common name	Flower colour	Flowering season	Remarks
1.	*Adenocalymma comosum*	Yellow trumpet vine	Yellow, large, scented	Mar - April	It is a heavy but slow growing evergreen climber. It is suitable on arch, wall. It grows in sun & semi-shade. Propagated by cuttings
2.	*Amphilophium mutisii*	-	Upper part deep violet, lower part pinkish white	Oct - Nov	It is evergreen climber. It forms a neat cover on pergola. Propagated by cutting & layering.

3.	*Anemopaegma chamberlaynii*	Trumpet vine	Light yellow, funnel shaped	Summer & early rains	It is an evergreen climber. It is suitable for covering wall, trellis & arch. Propagated by cuttings
4.	*Arrabidaea magnifica*	Glow vine, Purple funnel vine	Purplish violet funnel shaped	Summer	A shrubby climber with branches compressed when young It can be grown as a shrub Propagated by cuttings
5.	*Campsis grandiflora*	Trumpet flower	Deep orange	-	It is a shrubby climber. Flowers are widening trumpet like from a funnel shaped tube.
6.	*Campsis radicans*	Trumpet creeper, Humming bird vine	Orange	Summer & rainy	A scandent, deciduous plant. Propagated by cuttings. It is useful to climb on bare walls.
7.	*Clytostoma callistegioides*	Lavender trumpet vine,	yellow	Summer	An evergreen shrub. It is a showy ornamental species. It is suited for training over trellis (or) fence
8.	*Clytostoma purpureum*	-	Rosy-purple	Summer & rainy	It is an evergreen climber with slender branches. It should be grown on trellis (or) low walls. Propagated by cutting (or) layering

9.	*Doxantha unguis-cati*	Cats claw creeper	Yellow white deeper yellow lines in the throat	-	A slender, evergreen but extensively growing climber. It is a quick growing climber for covering greenhouse. It is used for light screening. Propagated by seeds (Or) layering
10.	*Phaedranthus buccinatoria*	Mexican blood flower	Crimson – red with scarlet sheen	Hot months	It is a large, woody, evergreen climber. It is ideal for training on wall, pergola (or) any high & strong support. It responds to pruning. Propagated by cuttings
11.	*Phaedranthus buccinatorius*	Scarlet trumpet vine	Crimson – red with scarlet sheen	Hot months	It is a large, woody, evergreen climber. It is ideal for training on wall, pergola (or) any high & strong support. It responds to pruning. Propagated by cuttings
12.	*Pithecoctenium cynanchoides*	Monkey comb	White	Hot months	It is a light climber. It prefers full sun & semi-shade. Propagated by cutting (or) layering.
13.	*Pyrostegia venusta*	Golden shower	Golden yellow	Jan - April	A popular deciduous climber climbing extensively by means of tendrils. It grows well in full sun (or)

					fence, trellis, wall, arch and porch. Propagated by cutting. Dry and thin shoots are to be removed after flowering

37. Climbing plant which belongs to the family Caesalpiniaceae are

S. No.	Botanical Name	Common name	Flower colour	Flowering season	Remarks
1.	*Bauhinia vahlii*	Maloo creeper, Mandari	White becoming yellowish	Mar - May	It is a vigorous & extensive climber, growing up to 15m (or) more. Multiplied from seeds (or) cuttings
2.	*Camoensia maxima*	Climbing camoensia	Large white tinged gold & fringed at the edges of the petals	Hot months	A vigorous, evergreen, shrubby climber. Suitable for climb over pergola, high wall. Propagated by cutting & layering
3.	*Clitoria ternatea*	Mussel-shell climber, Sangupushpam	-	-	An annual, twining climber. It is a quick growing climber suitable to train on low trellis, wall (or) fence. It is a perennial plant but grown as annual
4.	*Derris scandens*	Jewel vine	Pinkish white	Mar - Aug	A vigorous, evergreen, woody climber. It is a heavy climber, requires strong support. It is

					a good climber on pergola. It is used for screening. Propagated by seeds, cuttings & layering
5.	*Dioclea glycinoides*	-	-	Summer	A medium climber suitable to grow on low trellis. Propagated by seed & cuttings
6.	*Lathyrus odoratus*	Sweet pea	Blue, pink, maroon, purple, violet & different shades of red	-	It is one of the finest flowering. It is grown throughout the world. It grows well on trellis, fence made wire netting (or) bamboo splint. It prefers full sun. Propagated by seeds
7.	*Wisteria sinensis*	Chinese wisteria	Blue-violet	Hot months	A large, woody, spreading climber. Suitable to grow on arch, wall, arbour and trellis

38. Climbing plant which belongs to the family Capparidaceae is

S. No.	Botanical Name	Common name	Family	Flower colour	Flowering season	Remarks
1.	*Capparis horrida*	Caper berry	Capparidaceae	Numerous stamens white & turning to deep purple	Hot months	It is a spiny climber with extensive growth. It is a huge climberIt forms a thick cover on high wall (or) fence. It is propagated by layering

39. Climbing plant which belongs to the family Combretaceae are

S. No.	Botanical Name	Common name	Flower colour	Flowering season	Remarks
1.	*Combretum coccineum*	Flame vine	Orange red	Nov - May	A large, extensive shrubby climber. It is suitable to grow on low wall. Propagated by cuttings. It is thinned & pruned almost every year after flowering. It should be pruned every year
2.	*Combretum comosum*	-	Brilliant scarlet	Feb - June	A climbing shrub. It can trained on wall & pergola. Propagated by cutting (or) layering. It should be pruned every year
3.	*Combretum decandrum*	-	Orange - red	Summer	It is an extensively woody climber. It needs strong support & grown in large public gardens. Propagated by cutting (cr) layering
4.	*Quisqualis densiflora*	-	White, pink, red	-	It is an extensive climber
5.	*Quisqualis indica*	Rangoon creeper	White, turn pink & finally red	Summer & rainy	A large, vigorous, woody climber. It has quick growing habitIt needs strong support. Propagated by root suckers, cuttings & layers.

6.	*Quisqualis malabarica*	Malabar Madhu Malati	Open as white & change to pale red	-	It is used as a climber & also adapted to cultivation as bush. Propagated by sucker & layering

40. Climbing plant which belongs to the family Compositae are

S. No.	Botanical Name	Common name	Flower colour	Flowering season	Remarks
1.	*Senecio confusus*	Mexican daisy	-	Winter to Summer & also in rainy season	A semi-woody, evergreen climber. It is an attractive & useful climber for ground cover, on low wall (or) in pot with suitable support. Propagated from cutting
2.	*Vernonia elaeagnifolia*	Curtain creeper	White when young then it turns to pinkish white	-	A quick growing evergreen climber. It is a very useful climber for screening. Propagated by cuttings.

41. Climbing plant which belongs to the family Convolvulaceae are

S. No.	Botanical Name	Common name	Flowering season	Flower colour	Remarks
1.	*Argyreia nervosa*	Woolly morning glory/ Elephant creeper	Funnel – shaped, rosy purple	Year around flowering	A climbing shrub with white latex. It is a quick growing climber. Propagated by cuttings.

2.	*Ipomoea bonanox*	Moon flower	Large, greenish white, tubular	Summer & rains	It is a strong, woody climber of extensive growth. A quick growing, extensive climber. It is propagated by seed & cutting
3.	*Ipomoea horsfalliae*	Cardinal Creeper, Lady Doorly's Morning Glory	Magenta crimson	Throughout the year	It is a woody, spreading climber. Suitable to grow on wall, arch, trellis & porch. Propagated by cutting (or) layering. Pruning should be done during rainy season
4.	*Ipomoea learii*	Blue dawn flower	Large, bell shaped, blue-violet	Summer & rains	It is a quick growing, evergreen climber. It covers the wall uniformly. It prefers sunny locations. It is propagated by cuttings
5.	*Ipomoea palmata*	Railway creeper	Purple	Summer & rainy	A perennial, vigorous climber It is grown for the purpose of screening. Easily propagated by cutting
6.	*Ipomoea pes-caprae*	Railroad vine, Goat's foot vine	Purple	summer	Extensive climber. Flowers bell shaped. A quick growing climber for ground cover & screening
7.	*Ipomoea purpurea*	Common morning glory	Light blue (or) purple	-	A quick growing annual climber

8.	*Ipomoea rubro-caerulea*	Morning glory	Azure blue (or) purple	-	A quick growing annual climber. It is suitable to grow for covering arbours, wall and for screening
9.	*Ipomoea tuberosa*	Wood rose	Yellow funnel shaped	Hot months	A tuberous rooted climber rapidly forming extensive growth. Suitable for growing in sun & semi-shade on a strong support. Propagated by cuttings
10.	*Jacquemontia violacea*	-	Violet - blue	Throughout the year	A handsome, light climber It is very suitable for screening It prefers sun & semi-shade Propagated by cuttings
11.	*Porana paniculata*	Bridal creeper	White, small	Hot months	A strong shrubby climber. It is grown on a strong support. It prefers sun (or) semi-shade It needs pruning every year in the rainy season. Propagated by cutting & layering
12.	*Quamoclit coccinea*	Star ipomea	Crimson (or) orange	-	An annual, weak, glabrous twine. Easily propagated from seeds. It grows over trellis & low walls
13.	*Quamoclit lobata*	-	Crimson fading to pale yellow	Winter & early spring	An annual climber. It is a vigorous & quick growing. It grows well in pot & in ground

					on suitable support Propagated from seeds
14.	*Quamoclit pinnata*	Cypree vine	Scarlet, tubular	Summer & rains	It is an annual climber. It is suitable for growing on trellis. Propagated from seeds
15.	*Stictocardia tilifolia*	-	Bright crimson with yellow throat & scarlet red lines from the centre	Hot months	It is one of the most colourful climber. Suitable to grow on wall, porch, pergola. Propagated by cuttings

42. Climbing plant which belongs to the family Dilleniaceae are

S. No.	Botanical Name	Common name	Flower colour	Flowering season	Remarks
1.	*Tetracera sarmentosa*	-	White scented	summer	It is a large, evergreen, woody climber. It needs strong support & good sunshine. Propagated from cutting & layering
2.	*Thunbergia alata*	Black-eyed susan	Sulphur yellow	Cold months	It is a quick growing climber It is suitable for growing on low wall, trellis, pots with suitable support Propagated from seeds
3.	*Thunbergia fragrans*	Sweet Clock-Vine, White Lady	white	Winter to rains	A slender, perennial climber Propagated from cuttings

4.	*Thunbergia grandiflora*	Bengal Clock Vine, Bengal Trumpet Vine, Blue Sky Flower, Blue Sky Vine, Blue Trumpet Vine	Bright blue	Mar - Oct	A large, woody climber grown on pergola, fence (or) wall & also used for screeningIt is a heavy, spreading, evergreen climber requires strong support. Propagated by cuttings. It responds to pruning
5.	*Thunbergia laurifolia*	Laurel Clock Vine, laurel-leaved clockvine, laurel-leaved thunbergia, purple allamanda	Pale blue, white (or) yellow in the throat	Cold months	A large climber. Suitable for growing over walls & strong trellis. Propagated by seeds & cuttings
6.	*Thunbergia mysorensis*	Mysore Clock Vine, Dolls shoes, Brick and butter vine	Purple	Many months	A climbing shrub with long slender branches. A large climber suitable for growing over walls & strong trellis. Propagated from seed & cutting

43. Climbing plant which belongs to the family Liliaceae is

S. No.	Botanical Name	Common name	Flower colour	Flowering season	Remarks
1.	*Gloriosa rothschildiana*	Glory lily	Crimson, a dark purple mark near the base	Summer & rains	A beautiful climber

44. Climbing plant which belongs to the family Malpighiaceae are

S. No.	Botanical Name	Common name	Flower colour	Flowering season	Remarks
1.	*Banisteria laevifolia*	-	Bright yellow	Summer	A shrubby, slow growing climber. A good climber for wall, pergola & similar other supports. Propagated by cuttings
2.	*Hiptage madablota*	Hiptage, Helicopter Flower	White & yellow petals – silky outside & sweet scented	Feb - March	A large, evergreen, climbing shrub. It prefers sunny location. Propagated by cutting from green wood. The woody climber needs strong support for training.
3.	*Stigmaphyllon ciliatum*	Brazilian Golden Vine, Orchid Vine, Butterfly Vine, Amazon Vine	Golden yellow	Summer & rainy	It is an evergreen, medium climber with attractive foliage. Propagated by layers
4.	*Stigmaphyllum periplocifolium*	-	Yellow	Throughout the year	A quick growing climber. It should trained on a low & light support
5.	*Tristellateia australis*	Australian golden creeper	Yellow with a red filament	Throughout the year	A beautiful light climber with prominent lenticels. It is an ideal, light, evergreen climber It flowers profusely in pot on support Propagated by terminal cuttings

6.	*Vallaris heynei*	Bread flowe	White, fragrant	Feb - April	A heavy climber with woody branches. Pruning restricts the spread of the plant. It is propagated by cuttings

45. Climbing plant which belongs to the family Moraceae are

S. No.	Botanical Name	Common name	Flower colour	Flowering season	Remarks
1.	*Ficus pumila*	Creeping fig	-	-	A woody, ivy-like evergreen climber. It is commonly used for covering walls. It is propagated from cuttings
2.	*Ficus radicans*	Rooting fig	-	-	A trailing type with slender branches
3.	*Ficus villosa*	Shaggy fig	-	-	Creeping & rooting like ivy

46. Climbing plant which belongs to the family Nyctaginaceae are

S. No.	Botanical Name	Common name	Flower colour	Flowering season	Remarks
1.	*Bougainvillea buttiana*	-	-	-	Perianth tube is constricted at the middle
2.	*Bougainvillea glabra*	-	-	-	Perianth tube is swollen at the middle
3.	*Bougainvillea peruvinana*	-	-	-	Perianth is tubular & slender

4.	*Bougainvillea spectabilis*	-	-	-	Bracts ovate (or) elliptic, cordate at base
5.	*Bougainvillea spectoglabra*	-	-	-	It is very useful pot plant. It can be trained as bonsai It is hardy plant It is grown an arch, wall, porch & on any suitable support. It is an excellent plant for landscaping in dry & rockery areas and undulated sites

47. Climbing plant which belongs to the family Oleaceae are

S. No.	Botanical Name	Common name	Flower colour	Flowering season	Remarks
1.	*Jasminum angustifolium*	Wild jasmine	-	white, star-like flowers are exquisitely fragrant	-
2.	*Jasminum auriculatum*	Juhi	-	White, Flowers are fragrant, in many-flowered cymes	-
3.	*Jasminum flexile*	Taruni, River Jasmine, Irampanam mullai	-	White	-

4.	*Jasminum grandiflorum*	Royal Jasmine, Spanish jasmine	-	-	-
5.	*Jasminum humile*	Yellow Jasmine, Italian Jasmine		Yellow	
6.	*Jasminum nitidum*	Angelwing Jasmine, Royal jasmine, Shining jasmine, Star jasmine, Windmill jasmine	-	-	-
7.	*Jasminum officinale*	True jasmine, Poet's jasmine	-	-	-
8.	*Jasminum pubescence*	Star jasmine, Downy jasmine			
9.	*Jasminum sessiliflorum*	-	-	-	-
10.	*Jasminum trinerve*	-	-	-	-

48. Climbing plant which belongs to the family Polemoniaceae is

S. No.	Botanical Name	Common name	Flower colour	Flowering season	Remarks
1.	*Cobaea scandens*	Cup and saucer vine, Purple bell climber	Violet like calyx & cup like corolla firsh greenish changing to violet purple	-	It is a quick growing climber. It prefers warm humid condition. It is an annual. In mild climate, it is grown as perennial

49. Climbing plant which belongs to the family Polygonaceae are

S. No.	Botanical Name	Common name	Flower colour	Flowering season	Remarks
1.	*Antigonon guatemalense*	-	-	-	-
2.	*Antigonon leptopus*	Coral vine, Honolulu creeper, Mexican creeper, Bride's tears, Chain-of-love, Hearts on a chain, Love-vine	Rosy red	Hot months	A tuberous rooted, quick growing climber. It is grown in the garden mainly for the purpose of screening Propagated by cutting (or) ground layering

50. Climbing plant which belongs to the family Ranunculaceae are

No.	Botanical Name	Common name	Flower colour	Flowering season	Remarks
1.	*Clematis flammula*	Fragrant Virgin's Bower, fragrant clematis, sweet-scented virgin's bower	Pure white, fragrant	Summer & rainy	A slender climber. It is grown on wall, trellis (or) arbour. It prefers full sun (or) partial shade. Propagated by cuttings (or) layering
2.	*Clematis gouriana*	Gourian Clematis, Indian travelers's joy	White, fragrant	June - Sep	It is a quick growing, deciduous climber. It prefers bright sun & semi-shade

51. Climbing plant which belongs to the family Sapindaceae is

S. No.	Botanical Name	Common name	Flower colour	Flowering season	Remarks
1.	*Cardiospermum halicacabum*	Balloon Vine, Love-in-a-puff, heart pea, heartseed	Small white	-	A tendril climber, perennial but often grown as annual from seeds. It is growing to a height of about 3-4m. It is a quick growing climber. It is often grown in pots

52. Climbing plant which belongs to the family Solanaceae are

S. No.	Botanical Name	Common name	Flower colour	Flowering season	Remarks
1.	*Solandra nitida*	Cup of gold	Yellow with purplish stripes	-	A large shrubby climber It has an extensive growth. It needs strong support. It prefers full sun Propagated by cuttings
2.	*Solanum jasminoides*	Potato vine	Bluish white	Mar - Oct	A slender, shrubby climber. It grows well in partial shade. Propagated by cuttings
3.	*Solanum seaforthianum*	Potato creeper	Light purple (or) blue	June - Sep	A slender, slightly woody climber. It is an evergreen, light climber. It prefers semi-shade

4.	*Solanum wendlandii*	Giant potato vine	Lilac blue	Summer & rains	A large climbing shrub with few prickles on stems, branches & petioles. It prefers sun & semi-shade. It grows well on arch, wall, porch & strong trellis. Propagated by cutting

53. Climbing plant which belongs to the family Verbenaceae are

S. No.	Botanical Name	Common name	Flower colour	Flowering season	Remarks
1.	*Clerodendrum splendens*	Flaming Glorybower, Pagoda Flower, Bleeding Heart Vine	Scarlet flowers	Winter	It is a quick growing heavy climber. It is very suitable on arches, pergolas, walls (or) any other strong support
2.	*Clerodendrum thompsonae*	Bleeding heart	Sepals white, petals scarlet	Throughout year	It grows well in pots & in ground on suitable support in partial shade. Propagated by cutting & layering
3.	*Congea tomentosa*	Wooly Congea, Shower Orchid, or Shower of Orchid	White (or) pinkish mauve brack	Dec - March	It is woody, heavy climber. It is a quick growing, heavy climber. It requires strong support. It is propagated by cutting & layering
4.	*Faradaya splendida*	Glory vine	White	Summer	A woody climber of extensive growth. It is an

					large evergreen, quick growing climber. Suitable to grow on pergolas. Propagated by cutting & layering
5.	*Gmelina hystrix*	Parrot's Beak, Hedgehog	Bract – large greenish brown, corolla funnel shaped & yellow	April - Sep	Propagated from cuttings
6.	*Holmskioldia sanguinea*	Chinese-hat plant	Reddish orange	Throughout the year	A straggling, shrubby climber. The plant has an untidy growth. Pruning should be done in early rains
7.	*Petrea volubilis*	Purple wreath	Blue	Hot months	It is a large, woody vine with grayish bark. It is often growing on a porch, arch, pergola & wall. It prefers full sun. The petals drop off quickly leaving the lilac sepalsIt responds to pruning. Propagated by cutting

54. Climbing plant which belongs to the family Vitaceae is

S. No.	Botanical Name	Common name	Flower colour	Flowering season	Remarks
1.	*Cissus discolor*	Climbing Begonia, Rex Begonia Vine, Painted Cissus	Small yellow	-	It is a colourful foliage climber. It requires shady place. It is grown in green house also. Propagated by cutting

II. Fill in the blanks

1. A climber will climbs up with the help of sticky substances from growing points is ....
2. A Climber which climbs up with the help of hookers is ....
3. Common name of *Hoya carnosa* is....................
4. Planting distance for *Vernonia elaeagnifolia* is ....
5. Planting distance for *Ficus pumila* is ....

III. Choose the best answer

1. Climber which produce ornamental fruits is....................

   Bignonia unguiscati *Hedera nepalensis* *Clerodendrum inerme*
2. Climbers which is used for making topiary is....................

   Hedera nepalensis *Clerodendrum inerme* *Hiptage benghalensis*
3. Perennial climber is ....

   Thunbergia alata *Quamoclit pinnata* *Clematis* spp.
4. *Vallaris heynei* is a ....climber

   Heavy Light Medium
5. Climber which is grown for hedging is ....

   Jasminum grandiflorum *Solanum jasminoides* *Bougainvillea* spp.
6. Climber which produce fragrant flowers is ....

   Hedera nepalensis *Clerodendrum inerme* *Hiptage benghalensis*
7. Climber which is grown for low wall and trellis is....................

   Bougainvillea spp. *Clerodendrum thomsonae* *Jasminum grandiflorum*
8. Climber which is grown for indoor decoration is....................

   Monstera deliciosa *Ipomea horsfalliae* *Pyrostegia venusta*
9. A Climber which climbs up with the help of prickle is ....

   Passiflora caerulea *Rosa* spp *Combretum comosum*
10. Foliage climber is....................

    Cissus discolor *Clerodendrum inerme* *Hiptage benghalensis*

IV. Match the following

1.	Indian Ivy	:	*Clitoria ternata*
2.	Railway creeper	:	*Ipomea rubro-caerulea*
3.	Cup and saucer vine	:	*Lonicera japonica*
4.	butterfly pea creeper	:	*Ficus pumila*
5.	morning glory	:	*Ipomea palmata*
6.	Japanese honeysuckle	:	*Cobaea scandens*
7.	Golden shower	:	*Trachelospermum jasminoides*
8.	Bridal creeper	:	*Pyrostegia venusta*
9.	Black-eye susan	:	*Porana paniculata*
10.	Star jasmine	:	*Thunbergia alata*

Answers

II. Fill in the blanks

1. *Ficus repens*
2. *Combretum comosum*
3. Wax plant
4. 15 – 20 cm
5. 8 – 10 cm

III. Choose the best answer

1. *Hedera nepalensis*
2. *Clerodendrum inerme*
3. *Clematis* spp.
4. heavy
5. *Bougainvillea* spp.
6. *Hiptage benghalensis*
7. *Clerodendrum thomsonae*
8. *Monstera deliciosa*
9. *Rosa* spp
10. *Cissus discolor*

IV. Match the following

1. *Ficus pumila*
2. *Ipomea palmata*
3. *Cobaea scandens*
4. *Clitoria ternata*
5. *Ipomea rubro-caerulea*
6. *Lonicera japonica*
7. *Pyrostegia venusta*
8. *Porana paniculata*
9. *Thunbergia alata*
10. *Trachelospermum jasminoides*

ANNEXURE III

CLIMBERS AND CREEPERS

S. No.	Botanical Name	Common name	Family	Flower colour	Flowering time	Remarks
1.	*Adnenocalymma comosum*	-	Bignoniaceae	Yellow, large, scented	Mar - April	It is a heavy but slow growing evergreen climber. It is suitable on arch, wall. It grows in sun & semi-shade. Propagated by cuttings.
2.	*Allamanda cathartica*	Yellow allamanda	Apocynaceae	Yellow, large	Summer & Rainy	It prefer sunny location. It can be trained on wall, porch, pillar & arch. Propagated by cuttings
3.	*Allamanda violacea*	Purple allamanda	Apocynaceae	Light violet, large	Summer & Rainy	A scrambling shrub with terete branches. It prefers a sunny location. It can be trained on low wall (or) pillar. Propagated by cuttings
4.	*Amphilophium mutisii*	-	Bignoniaceae	Upper part deep violet, lower part pinkish white	Oct - Nov	It is evergreen climber. It forms a neat cover on pergola. Propagated by cutting & layering
5.	*Anemopaegma chamberlaynii*	-	Bignoniaceae	Light yellow, funnel shaped	Summer & early rains	It is an evergreen climber. It is suitable for covering wall, trellis & arch. Propagated by cuttings

6.	*Antigonon guatemalense*	-	Polygonaceae	-	-	-
7.	*Antigonon leptopus*	-	Polygonaceae	Rosy red	Hot months	A tuberous rooted, quick growing climber. It is grown in the garden mainly for the purpose of screening. Propagated by cutting (or) ground layering
8.	*Aristolochia elegans*	Calico flower	Aristoloc-hiaceae	Corolla tube is yellow – green, flower is purple & white – blotched, white on the exterior, yellow eye	Hot months	Curious pouch flower. It is an ideal light climber. Propagated from cuttings
9.	*Aristolochia grandiflora*	Pelican flower/ swan flower	Aristoloc-hiaceae	Large flowers offensive odour, purple – blotched & veined	Oct - Nov	A shrubby, extensive climber. It is a heavy climber, needs strong support. It should be grown away from house. Propagated by seeds, cutting & layering
10.	*Aristolochia ornithocephala*	The bird's head flower	Aristoloc-hiaceae	Netted (or) spotted dark purple, tube purple, upper lip violet inside, pouch – like,	-	It is a free flowering climber. It is grown on walls & other strong supports. It prefers sunny locations. Propagated from cuttings of matured wood (or) by seed

				often shining dark purple		
11.	*Aristolochia ringens*	-	Aristoloc-hiaceae	Pale green in colour, hairy within	Mar – May & also in Oct - Dec	It is a large evergreen climber. It is a quick growing climber. It is suitable for growing on wall.
12.	*Aristolochia tomentosa*	-	Aristoloc-hiaceae	Yellow (or) yellowish green	-	A tall climber. It is quite hardy, grows to a great height. It is useful for the purpose of screening.
13.	*Argyreia nervosa*	Woolly morning glory/Elephant creeper	Convolvu-laceae	Funnel – shaped, rosy purple	Year around flowering	A climbing shrub with white latex. It is a quick growing climber. Propagated by cuttings
14.	*Arrabidaea magnifica*	-	Bignoniaceae	Purplish violet funnel shaped	Summer	A shrubby climber with branches compressed when young. It can be grown as a shrub. Propagated by cuttings
15.	*Artabotrys odoratissimus*	Tail grape/ Climbing ylangylang	Annonaceae	Greenish yellow, strong fragrance	Hot & rainy season	It is an evergreen, large, shrubby climber. Propagated by seeds (or) layers
16.	*Banisteria laevifolia*	-	Malpighi-aceae	Bright yellow	Summer	A shrubby, slow growing climber. A good climber for wall, pergola & similar other supports. Propagated by cuttings

17.	*Bauhinia vahlii*	-	Caesalpiniaceae	White becoming yellowish	Mar - May	It is a vigorous & extensive climber, growing up to 15m (or) more. Multiplied from seeds (or) cuttings.
18.	*Beaumontia grandiflora*	Nepal trumpet	Apocynaceae	Large, white	Jan - March	It is a heavy, evergreen climber. It is a quick growing. It flowers well in both sun & semi-shade. Propagated by cutting & layering
19.	*Bougainvillea buttiana*	-	Nyctaginaceae	-	-	Perianth tube is constricted at the middle
20.	*Bougainvillea glabra*	-	Nyctaginaceae	-	-	Perianth tube is swollen at the middle
21.	*Bougainvillea peruvinana*	-	Nyctaginaceae	-	-	Perianth is tubular & slender
22.	*Bougainvillea spectabilis*	-	Nyctaginaceae	-	-	Bracts ovate (or) elliptic, cordate at base
23.	*Bougainvillea spectoglabra*	-	Nyctaginaceae	-	-	It is very useful pot plant. It can be trained as bonsai. It is hardy plant. It is grown an arch, wall, porch & on any suitable support. It is an excellent plant for landscaping in dry & rockery areas and undulated sites

24.	*Camoensia maxima*	-	Fabaceae	Large white tinged gold & fringed at the edges of the petals	Hot months	A vigorous, evergreen, shrubby climber. Suitable for climb over pergola, high wall. Propagated by cutting & layering.
25.	*Campsis grandiflora*	Trumpet flower	Bignoniaceae	Deep orange	-	It is a shrubby climber. Flowers are widening trumpet like from a funnel shaped tube.
26.	*Campsis radicans*	-	Bignoniaceae	Orange	Summer & rainy	A scandent, deciduous plant Propagated by cuttings. It is useful to climb on bare walls
27.	*Capparis horrida*	-	Cappari-daceae	Numerous stamens white & turning to deep purple	Hot months	It is a spiny climber with extensive growth. It is a huge climber. It forms a thick cover on high wall (or) fence. It is propagated by layering
28.	*Cardiospermum halicabum*	-	Sapindaceae	Small white	-	A tendril climber, perennial but often grown as annual from seeds. It is growing to a height of about 3-4m. It is a quick growing climber. It is often grown in pots
29.	*Chonemorpha macrophylla*	-	Apocynaceae	White yellowish in the throat, fragrant	Summer & rainy	An evergreen, heavy climber. It is suitable to grow on pergola or other strong

						support. Propagated by cutting & layering
30.	*Cissus discolor*	-	Vitaceae	Small yellow	-	It is a colourful foliage climber. It requires shady place. It is grown in green house also. Propagated by cutting
31.	*Clematis flammula*	-	Ranuncu-laceae	Pure white, fragrant	Summer & rainy	A slender climber. It is grown on wall, trellis (or) arbour. It prefers full sun (or) partial shade. Propagated by cuttings (or) layering
32.	*Clematis gouriana*	-	Ranuncu-laceae	White, fragrant	June - Sep	It is a quick growing, deciduous climber. It prefers bright sun & semi-shade
33.	*Clerodendrum splendens*	-	Verbenaceae	Scarlet flowers	Winter	It is a quick growing heavy climber. It is very suitable on arches, pergolas, walls (or) any other strong support
34.	*Clerodendrum thompsonae*	Bleeding heart	Verbenaceae	Sepals white, petals scarlet	Throughout year	It grows well in pots & in ground on suitable support in partial shade. Propagated by cutting & layering
35.	*Clitoriaternatea*	Mussel-shell climber	Fabaceae	-	-	An annual, twining climber. It is a quick growing climber suitable to train on low trellis,

						wall (or) fence. It is a perennial plant but grown as annual
36.	*Clytostoma callistegioides*	-	Bignoniaceae	yellow	Summer	An evergreen shrub. It is a showy ornamental species. It is suited for training over trellis (or) fence
37.	*Clytostoma purpureum*	-	Bignoniaceae	Rosy-purple	Summer & rainy	It is an evergreen climber with slender branches. It should be grown on trellis (or) low walls. Propagated by cutting (or) layering
38.	*Cobaea scandens*	Cup and saucer vine	Polemoni-aceae	Violet like calyx & cup like corolla firsh greenish changing to violet purple	-	It is a quick growing climber. It prefers warm humid condition. It is an annual. In mild climate, it is grown as perennial
39.	*Combertum coccineum*	-	Combretaceae	Orange red	Nov - May	A large, extensive shrubby climber. It is suitable to grow on low wall. Propagated by cuttings. It is thinned & pruned almost every year after flowering. It should be pruned every year
40.	*Combertum comosum*	-	Combretaceae	Brilliant scarlet	Feb - June	A climbing shrub. It can trained on wall & pergola.

						Propagated by cutting (or) layering. It should be pruned every year
41.	*Combertum decandrum*	-	Combretaceae	Orange - red	Summer	It is an extensively woody climber. It needs strong support & grown in large public gardens. Propagated by cutting (or) layering
42.	*Congea tomentosa*	-	Verbenaceae	White (or) pinkish mauve brack	Dec - March	It is woody, heavy climber. It is a quick growing, heavy climber. It requires strong support. It is propagated by cutting & layering
43.	*Cryptostegia grandiflora*	Indian rubber vine	Asclepiada-ceae	Reddish purple turning lilac	Mar - Sep	A woody, evergreen plant contains white latex. Propagated by cuttings. It responds to pruning
44.	*Cryptostegia madagascariensis*	-	Asclepiad-aceae	-	-	-
45.	*Derris scandens*	-	Fabaceae	Pinkish white	Mar - Aug	A vigorous, evergreen, woody climber. It is a heavy climber, requires strong support. It is a good climber on pergola. It is used for screening. Propagated by seeds, cuttings & layering

46.	*Dioclea glycinoides*	-	Fabaceae	-	Summer	A medium climber suitable to grow on low trellis. Propagated by seed & cuttings
47.	*Doxantha unguis-cati*	-	Bignoniaceae	Yellow white deeper yellow lines in the throat	-	A slender, evergreen but extensively growing climber. It is a quick growing climber for covering greenhouseIt is used for light screening. Propagated by seeds (Or) layering
48.	*Echites caryophyllata*	-	Apocynaceae	White, scented	Late summer & rainy	A heavy climber It grows well in semi-shade & full sun. Propagated by cutting & layeringIdeal climber on pergola, arch, wall & arbour in a sunny location
49.	*Echites rubro-venosa*	-	Apocynaceae	-	-	A slender climber. It grows well in partial sunlight, warm & moist conditions. It is a good plant for growing on pots with suitable support
50.	*Faradaya splendida*	-	Verbenaceae	white	Summer	A woody climber of extensive growth. It is an large evergreen, quick growing climber. Suitable to grow on

						pergolas. Propagated by cutting & layering
51.	*Ficus pumila*	Creeping fig	Moraceae	-	-	A woody, ivy-like evergreen climber. It is commonly used for covering walls. It is propagated from cuttings
52.	*Ficus radicans*	Rooting fig	Moraceae	-	-	A trailing type with slender branches
53.	*Ficus villosa*	Shaggy fig	Moraceae	-	-	Creeping & rooting like ivy
54.	*Gloriosa rot hschildiana*	Glory lily	Liliaceae	Crimson, a dark purple mark near the base	Summer & rains	A beautiful climber
55.	*Gmelina hystrix*	-	Verbenaceae	Bract – large greenish brown, corolla funnel shaped & yellow	April - Sep	Propagated from cuttings
56.	*Hiptage madablota*	-	Malpighi-aceae	White & yellow petals – silky outside & sweet scented	Feb - March	A large, evergreen, climbing shrub. It prefers sunny location. Propagated by cutting from green wood. The woody climber needs strong support for training
57.	*Holmskioldia sanguinea*	Chinese-hat plant	Verbenaceae	Reddish orange	Throughout the year	A straggling, shrub by climber. The plant has an untidy growth. Pruning should be done in early rains

58.	*Hoya bella*	Miniature wax plant	Asclepiad-aceae	Waxy white with purple centre	Summer	A slender, bushy plant with drooping branches. A small attractive plant. It can be grown as shrub (or) climber with light support. It is grown in pots
59.	*Hoya carnosa*	Wax plant	Asclepia-daceae	Pinkish white with a red, star shaped crown, fragrant	Hot months	A vigorous growing twines attached to support by means of roots. It prefers partial shade It grows well in pot on suitable support. Propagated by cutting
60.	*Hoya imperialis*	Honey plant	Asclepia-daceae	Large, waxy dull - pink	Hot months	A tall, spreading climber. It produces cluster of beautiful flowers. Propagated by cuttings
61.	*Ipomoea bonanox*	Moon flower	Convolvu-laceae	Large, greenish white, tubular	Summer & rains	It is a strong, woody climber of extensive growth. A quick growing, extensive climber. It is propagated by seed & cutting
62.	*Ipomoea horsfalliae*	-	Convolvu-laceae	Magenta crimson	Throughout the year	It is a woody, spreading climber. Suitable to grow on wall, arch, trellis & porch. Propagated by cutting (or) layering. Pruning should be done during rainy season.

63.	*Ipomoea learii*	Blue dawn flower	Convolvu-laceae	Large, bell shaped, blue-violet	Summer & rains	It is a quick growing, evergreen climber. It covers the wall uniformly .It prefers sunny locations. It is propagated by cuttings
64.	*Ipomoea palmata*	Railway creeper	Convolvu-laceae	Purple	Summer & rainy	A perennial, vigorous climber. It is grown for the purpose of screening. Easily propagated by cutting
65.	*Ipomoea pes-caprae*	-	Convolvu-laceae	Purple	summer	Extensive climber. Flowers bell shaped. A quick growing climber for ground cover & screening
66.	*Ipomoea purpurea*	Common morning glory	Convolvu-laceae	Light blue (or) purple	-	A quick growing annual climber
67.	*Ipomoea rubro-caerulea*	Morning glory	Convolvu-laceae	Azure blue (or) purple	-	A quick growing annual climber. It is suitable to grow for covering arbours, wall and for screening
68.	*Ipomoea tuberosa*	Wood rose	Convolvu-laceae	Yellow funnel shaped	Hot months	A tuberous rooted climber rapidly forming extensive growth. Suitable for growing in sun & semi-shade on a strong support. Propagated by cuttings

69.	*Jacquemontia violacea*	-	Convolvu-laceae	Violet - blue	Throughout the year	A handsome, light climber. It is very suitable for screening. It prefers sun & semi-shade. Propagated by cuttings
70.	*Jasminum angustifolium*	-	Oleaceae	-	-	-
71.	*Jasminum auriculatum*	-	Oleaceae	-	-	-
72.	*Jasminum flexile*	-	Oleaceae	-	-	-
73.	*Jasminum grandiflorum*	Spanish jasmine	Oleaceae	-	-	-
74.	*Jasminum nitidum*	-	Oleaceae	-	-	-
75.	*Jasminum officinale*	White jasmine	Oleaceae	-	-	-
76.	*Jasminum sessiliflorum*	-	Oleaceae	-	-	-
77.	*Jasminum trinerve*	-	Oleaceae	-	-	-
78.	*Jasminum undulatum*	-	Oleaceae	-	-	-
79.	*Lathyrus odoratus*	Sweet pea	Fabaceae	Blue, pink, maroon, purple, violet & different shades of red	-	It is one of the finest flowering. It is grown throughout the world. It grows well on trellis, fence made wire netting (or) bamboo splint. It prefers full sun. Propagated by seeds

80.	*Lonicera confusa*	-	Caprifolia-ceae	White fading to yellow	Summer	An evergreen, twining shrub. It grows well on arch, trellis (or) low wall. Propagated from seed
81.	*Lonicera japonica*	Japanese honeysuckle	Caprifolia-ceae	First white gradually turning yellow (or) white tinged purple, fragrant	Summer & rain	A beautiful, evergreen climber. Suitable for mild climate. An ideal climber on low wall. It flowers well in sun & semi-shade. Propagated by cuttings.
82.	*Lonicera sempervirens*	Trumpet honeysuckle	Caprifolia-ceae	Orange scarlet, yellow inside	Summer	It is an extensive woody climber. It is best adopted for trellis (or) pergola
83.	*Melodinus monogynous*	-	Apocynaceae	White, scented	March - May	It grows well on strong arch & pergola. It prefers full sun. Propagated by cutting
84.	*Odontadenia grandiflora*	-	Apocynaceae	Bright yellow, large, scented	Throughout the year	It is a large, shrubby climber. It prefers full sun & semi-shade. It is often grown on wall & pergola. Propagated from cutting & layering
85.	*Pandorea jasminoides*	-	Bignonni-aceae	Rosy pink, almost purple in the throat	Summer & rains	It is an evergreen climber. Suitable for growing on low walls & trellis. Propagated by cutting (or) layering
86.	*Passiflora caerulea*	-	Passifloaceae	Pale pink, outer filaments	Rainy season	A large evergreen climber. It is a hardy climber. Suitable

				blue at the tip, white in the middle, purple at the base		for screening. Propagated from suckers, seeds, cutting & layering
87.	*Passiflora coccinea*	Red passion flower	Passifloaceae	White, often tinted purple, white in the upper half, violet (or) purple below	Summer	A strong, woody, quick growing climber. It covers fences & trellis. Propagation by seeds, cuttings & layering
88.	*Passiflora edulis*	Passion fruit	Passifloaceae	-	-	-
89.	*Passiflora laurifolia*	-	Passifloaceae	Sepals & petals green outside & red within	Summer & rains	A strong, glabrous quick growing. It covers arch, wall, trellis It prefers sun (or) semi-shade. Propagated by cuttings & layering
90.	*Passiflora loudonii*	-	Passifloaceae	-	-	-
91.	*Passiflora quadrangularis*	Giant granadilla	Passifloaceae	Pink (or) mauve outside, white, blue – violet (or) pinkish inside	-	A strong robust climber with quadrangular stem
92.	*Passiflora racemosa*	Princess charlotte's passion flower	Passifloaceae	scarlet	Summer & rain	A climber with cylindrical, glabrous, woody stem. It is suitable to grow on arch & pergolas
93.	*Passiflora raddina*	-	Passifloaceae	Scarlet	-	A slender, glabrous climber. Propagated by cuttings.

94.	*Passiflora violacea*	Purple passion flower	Passifloaceae	Purplish – blue, fragrant, sepals pinkish lilac, petals violet	Summer & rainy	A medium, evergreen climber
95.	*Pergularia odoratissimus*	-	Asclepiad-aceae	Greenish yellow, scented	Hot	It is an evergreen climber. It produces flowers resembling those of a cowslip. It needs pruning. Propagated by cuttings & layering
96.	*Petrea volubilis*	Purple wreath	Verbenaceae	Blue	Hot months	It is a large, woody vine with grayish bark. It is often growing on a porch, arch, pergola & wall. It prefers full sun. The petals drop off quickly leaving the lilac sepals. It responds to pruning. Propagated by cutting
97.	*Phaedranthus buccinatoria*	-	Bignoniaceae	Crimson – red with scarlet sheen	Hot months	It is a large, woody, evergreen climber. It is ideal for training on wall, pergola (or) any high & strong support. It responds to pruning. Propagated by cuttings.
98.	*Pithecoctenium cynchoides*	-	Bignoniaceae	White	Hot months	It is a light climber. It prefers full sun & semi-shade. Propagated by cutting (or) layering

99. *Porana paniculata*	Bridal creeper	Convolvu-laceae	White, small	Hot months	A strong shrubby climber. It is grown on a strong support. It prefers sun (or) semi-shade. It needs pruning every year in the rainy season. Propagated by cutting & layering
100. *Pseudocalymma alliaceum*	Garlic vine	Bignoniaceae	Mauve	Oct – Feb	An evergreen plant with dense foliage. It is a quick growing. Propagated by cuttings. It is an ideal climber on arch, porch, wall & pergola
101. *Pyrostegia venusta*	Golden shower	Bignoniaceae	Golden yellow	Jan - April	A popular deciduous climber climbing extensively by means of tendrils. It grows well in full sun (or) fence, trellis, wall, arch and porch. Propagated by cutting. Dry and thin shoots are to be removed after flowering
102. *Quamoclit coccinea*	Star ipomea	Convolvu-laceae	Crimson (or) orange	-	An annual, weak, glabrous twine Easily propagated from seeds.It grows over trellis & low walls
103. *Quamoclit lobata*	-	Convolvu-laceae	Crimson fading to pale yellow	Winter & early spring	An annual climber. It is a vigorous & quick growing. It

					grows well in pot & in ground on suitable support. Propagated from seeds
104. *Quamoclit pinnata*	Cypree vine	Convolvu-laceae	Scarlet, tubular	Summer & rains	It is an annual climber. It is suitable for growing on trellis. Propagated from seeds
105. *Quisqualis densiflora*	-	Combretaceae	White, pink, red	-	It is an extensive climber
106. *Quisqualis indica*	Rangoon creeper	Combretaceae	White, turn pink & finally red	Summer & rainy	A large, vigorous, woody climber. It has quick growing habit. It needs strong support. Propagated by root suckers, cuttings & layers
107. *Quisqualis malabarica*	-	Combretaceae	Open as white & change to pale red	-	It is used as a climber & also adapted to cultivation as bush. Propagated by sucker & layering
108. *Rhynchodia wallichii*	-	Apocynaceae	Yellowish white with a reddish centre	Summer & rains	A large, woody, evergreen climber. It grows well on pergola & wall. It serve for the purpose of screening. Propagated by cutting
109. *Senecio confusus*	Mexican daisy	Compositae	-	Winter to summer & also in rainy season	A semi-woody, evergreen climber. It is an attractive & useful climber for ground cover, on low wall (or) in pot with suitable support. Propagated from cutting.

110.	*Solandra nitida*	Cup of gold	Solanaceae	Yellow with purplish stripes	-	A large shrubby climber. It has an extensive growth. It needs strong support. It prefers full sun. Propagated by cuttings
111.	*Solanum jasminoides*	Potato vine	Solanaceae	Bluish white	Mar - Oct	A slender, shrubby climber. It grows well in partial shade. Propagated by cuttings
112.	*Solanum seaforthianum*	Potato creeper	Solanaceae	Light purple (or) blue	June - Sep	A slender, slightly woody climber. It is an evergreen, light climber. It prefers semi-shade
113.	*Solanum wendlandii*	Giant potato vine	Solanaceae	Lilac blue	Summer & rains	A large climbing shrub with few prickles on stems, branches & petioles. It prefers sun & semi-shade. It grows well on arch, wall, porch & strong trellis. Propagated by cutting
114.	*Stephanotis floribunda*	Madagascar jasmine	Asclepiad-aceae	Pure white, very fragrant	Summer & rainy	A slender climber. It is an ideal climber on low wall (or) trellis in pot. Propagated by cutting (or) layering
115.	*Stictocardia tilifolia*	-	Convolvul-aceae	Bright crimson with yellow throat & scarlet red lines from the centre	Hot months	It is one of the most colourful climber. Suitable to grow on wall, porch, pergola. Propagated by cuttings

116. *Stigmaphyllon ciliatum*	-	Malpighia-ceae	Golden yellow	Summer & rainy	It is an evergreen, medium climber with attractive foliage. Propagated by layers
117. *Stigmaphyllum periplocifolium*	-	Malpighia-ceae	Yellow	Throughout the year	A quick growing climber. It should trained on a low & light support
118. *Strophanthus grandiflorus*	-	Apocynaceae	Light yellow, streaked with red, funnel shaped	Summer months	Large, woody climber. It is an extensive evergreen climber. Suitable for large gardens
119. *Strophanthus gratus*		Apocynaceae	White, tinted purple, scented	Summer & rainy	A large evergreen woody climber. It requires strong support & enough space. It is suitable on wall, pergola. Propagated by cuttings
120. *Tetracera sarmentosa*	-	Dilleniaceae	White scented	summer	It is a large, evergreen, woody climber. It needs strong support & good sunshine. Propagated from cutting & layering
121. *Thunbergia alata*	Black-eyed susan	Dilleniaceae	Sulphur yellow	Cold months	It is a quick growing climber. It is suitable for growing on low wall, trellis, pots with suitable support. Propagated from seeds

122. *Thunbergia fragrans*	-	Dilleniaceae	White	Winter to rains	A slender, perennial climber. Propagated from cuttings
123. *Thunbergia grandiflora*	-	Dilleniaceae	Bright blue	Mar - Oct	A large, woody climber grown on pergola, fence (or) wall & also used for screening. It is a heavy, spreading, evergreen climber requires strong support. Propagated by cuttings. It responds to pruning
124. *Thunbergia laurifolia*	-	Dilleniaceae	Pale blue, white (or) yellow in the throat	Cold months	A large climber. Suitable for growing over walls & strong trellis. Propagated by seeds & cuttings
125. *Thunbergia mysorensis*	-	Dilleniaceae	Purple	Many months	A climbing shrub with long slender branches. A large climber suitable for growing over walls & strong trellis. Propagated from seed & cutting
126. *Trachelspermum jasminoides*	-	Apocynaceae	White small, fragrant	Hot months	A large evergreen climbing shrub with white latex. It prefers full sun & partial shade. It is suitable to grow on fence, wall, trellis. Propagated by cutting (or) layering.

127. *Tristellateia australis*	Australian golden creeper	Malpighi-aceae	Yellow with a red filament	Throughout the year	A beautiful light climber with prominent lenticels. It is an ideal, light, evergreen climber. It flowers profusely in pot on support. Propagated by terminal cuttings
128. *Vallaris heynei*	-	Malpighi-aceae	White, fragrant	Feb - April	A heavy climber with woody branches. Pruning restricts the spread of the plant. It is propagated by cuttings
129. *Vernonia elaeagnifolia*	-	Compositae	White when young then it turns to pinkish white	-	A quick growing evergreen climber. It is a very useful climber for screening. Propagated by cuttings
130. *Wisteria sinensis*	Chinese wisteria	Fabaceae	Blue-violet	Hot months	A large, woody, spreading climber. Suitable to grow on arch, wall, arbour and trellis

Chapter **7**

Annuals

I. General information

1. A group of ornamental plants which complete their life cycle in one season or in a year is called **annuals**
2. Time of sowing of summer annual seeds is **February- March**
3. Time of sowing of rainy season annual seeds is **June- July**
4. Time of sowing of winter annual seeds is **September- October**
5. Based on growing season, annuals are classified into **summer, rainy & winter**
6. Based on hardiness, annuals are classified into **hardy, half hardy & tender**
7. Based on photoperiodic requirement annuals are classified into **short day, long day, day neutral, long short day, short long day & intermediate**
8. **Summer season annuals** are

   Amaranthus - *Amaranthus* sp - Amaranthaceae

   Balsam –*Impatiens balsamina*- Balsaminaceae

   Cock's comb - *Celosia* sp - Amaranthaceae

   Gaillardia - *Gaillardia pulchella* - Asteraceae

   Gomphrena - *Gomphrena globosa* - Amaranthaceae

   Kochia- *Kochia scoparia* var. tricophylla - Chenopodiaceae

   Portulaca - *Portulaca grandiflora* - Portulacaceae

   Sun flower - *Helianthus annus* - Asteraceae

Tithonia –*Tithonia rotundifolia* – Asteraceae

Zinnia – *Zinnia angustifolia&Zinnia linearis* – Asteraceae

9. **Winter season annuals** are

Acroclinum – *Acroclinum roseum* - Asteraceae

Ageratum – *Ageratum houstonianum* – Asteraceae

Annual chrysanthemum – *Chrysanthemum coronarium, Chrysanthemum carinatum,* Chrysanthemum segetum – Asteraceae

Arctotis – *Arctotis stoechadifolia* - Asteraceae

Aster – *Aster amellus* - Asteraceae

Brachycome – *Brachycome iberidifolia* - Asteraceae

Calendula – *Calendula officinalis* - Asteraceae

Cineraria – *Senecio cruentus* - Asteraceae

Coreopsis – *Coreopsis tinctoria* - Asteraceae

Corn flower – *Centaurea cyanus* - Asteraceae

Cosmos – *Cosmos bipinnatus* - Asteraceae

Dahlia – *Dahlia variabilis* - Asteraceae

Daisy – *Bellis perennis* - Asteraceae

Dimorphotheca – *Dimorphotheca aurantiaca*- Asteraceae

Gamolepsis – *Gamolepsis tagetes* - Asteraceae

Gazania- *Gazania splendens* - Asteraceae

Gypsophila – *Gypsophila elegans* - Asteraceae

Saponaria – *Saponaria vaccaria* – Caryophyllaceae

Sweet William – *Dianthus barbatus* – Caryophyllaceae

Helichrysum – *Helichrysum bracteatum* – Asteraceae

Rudbeckia – *Rudbeckia bicolour* – Asteraceae

Sweet sultan –*Centaurea moschata*- Asteraceae

Venidium- *Venidium fastuosum*- Asteraceae

Anchusa- *Anchusa capensis* – Boraginaceae

Antirrhinum – *Antirrhinum majus* – Scrophulariaceae

Linaria - *Linaria bipartita* - Scrophulariaceae

Mimulus-*Mimulus tigrinus* - Scrophulariaceae

Nemasia - *Nemasia strumosa* - Scrophulariaceae

Bells of Ireland- *Molucella laevis* - Labiatae

Salvia - *Salvia splendens*- Labiatae

Californian poppy- *Eschscholzia californica*- Papavraceae

Shirley poppy- *Papaver rhoeas* - Papavaraceae

Candy tuft- *Iberis sp* –Cruciferae

Stock- *Mathiola incana*- Cruciferae

Sweet Alyssum- *Alyssum maritinum* - Cruciferae

Wall flower- *Cheiranthus cheiri* - Cruciferae

Clarkia - *Clarkia elegans*- Onagraceae

Clianthus- *Clianthus dampieri*- Leguminosae

Lupin- *Lupinus hartwegii*- Leguminosae

Sweet pea- *Lathyrus odoratus*- Leguminosae

Holly Hock- *Althea rosea* - Malvaceae

Iceplant- *Mesembryanthemum criniflorum* - Aizoaceae

Lady's lace - *Pimpinella monoica* - Umbelliferae

Larkspur- *Delphinium hybridum* - Ranuncullaceae

Nigella- *Nigella damascene* - Ranuncullaceae

Linum- *Linum grandiflorum* var rubrum - Linaceae

Mignonette- *Reseda odorata* - Resedaceae

Nasturtium- *Tropaeolum majus* - Tropaeolaceae

Pansy- *Viola wittorckiana*- Violaceae

Petunia- *Petunia hybrida*- Solanaceae

Schizanthus- *Schizanthus wisetonensis* - Solanaceae

Phlox- *Phlox drummondii* - Polemoniaceae

Statice- *Limonium sinuatum* - Plumbaginaceae

Verbena- *Verbena hybrida*- Verbenaceae

10. Examples for hardy annuals are **Digitalis, Rudbeckia & Viola**
11. Examples for half hardy annuals are **Antirrhinum, Stock**
12. Examples for tender annuals are **oxalis & pansy**
13. Annuals used as fragrant flowers are **Mignonette, Dianthus, sweet pea, sweet sultan, sweet William, sweet alyssum, stock**
14. Annuals used for hanging basket are **Daisy, Nasturtium, Verbana, Phlox, Sweet Alyssum, Portulaca, Torenia, Lobelia & Hymenantherum**
15. Annuals used as cut flower are **Carnation, Sweet pea, Sweet William, Aster, Larkspur, Helichrysum, Lupin, Antirrhinum, sweet sultan, corn flower**
16. Annuals used as loose flower are **Marigold, Annual Chrysanthemum, China Aster, Sunflower, Zinnia, Gaillardia, Celosia, Gomphrena, etc.**
17. Annuals used for shady situation are **Salvia & Cineraria**
18. Annuals used for rock garden are **Ice plant, Nasturtium, Verbena, Phlox, Gamolepsis**
19. Annuals used for screening purpose are **Holly hock & sweet pea**
20. Annuals used for peculiar shape is **dianthus**
21. Annuals used as pot plant are **Carnation, Antirrhinum, Aster & Petunia**
22. Annuals used as dry flowers are **Statice, Helichrysum, Acroclinum,Nigella, Lady's lace**
23. Types of herbaceous border are **single face & Double face**
24. A term used for planting of annuals in the border of a lawn or shrubbery in a definite pattern is **herbaceous border**
25. Ideal site for making herbaceous border is **open sunny, south or south- east direction**
26. Width of herbaceous border should be of **1-1.5 m**
27. Primary colours are **Red, yellow & Blue**
28. Secondary colours are **Orange, Green & Violet**
29. Tertiary colours are **Cream, Pink & shades of colours**

30. Colour of a plant which dominate throughout the year in the garden is **Green**
31. Colour of a flower which dominate during winter is **red**
32. Colour of a flower which dominate during spring is **yellow**
33. Warm colours are **red, yellow & orange**
34. Cool colours are **green & blue**
35. A colour scheme in which annuals are restricted with the use of single colour or its shades is called **Monochromatic**
36. A colour scheme in which annuals are arranged according to the nearest wave length of colours is **Analogous**
37. Analogous colour scheme is also called as **harmonious colour scheme**
38. Closely related colours are also known as **Analogous** colours
39. A colour scheme in which annuals are arranged by its two opposite colour is **Contrast**
40. Contrast colour scheme is also called as **complementary colour scheme**
41. Examples for opposite colours are **red-green, yellow-voilet, blue-orange**
42. Neutral colours are **white, black & gray**
43. Self pollinated annuals are **lupin, sweet pea & salvia**
44. Often cross pollinated annuals are **Antirrhinum, larkspur, linaria, phlox, pansy**
45. Both rainy & summer season annuals are **balsam, cocks comb, amaranthus, gaillardia, torenia, gomphrena, kochia, portulaca, sunflower & zinnia**
46. Annual seeds which germinate only in light are **Nicotiana, Lobelia & Echium**
47. Annual seeds which germinate only in dark **Nigella, Phlox, Amaranthus, Allium**
48. Short day annuals are **Amaranthus, Chrysanthemum, Cosmos & Salvia**
49. Long day annuals are **Antirrhinum, Dianthus, Petunia & Rudbeckia**

50. Day neutral annuals are **Gomphrena & Impatiens**
51. Intermediate day annual is **Coleus**
52. Long short day annual is **Aloe**
53. Short long day annual is **Campanula & white clover**
54. Leading annual flower seed producing states in India are **Punjab, Haryana, Himachal Pradesh, Jammu & Kashmir, Maharastra, Karnataka, West Bengal & Jharkhand**
55. A state which contributes 50% share in flower seed production in India is **Punjab**
56. Annuals which possess self- incompatibility are **Ageratum, Antirrhinum, Daisy, Gerbera, Nicotiana & Petunia**
57. Isolation distance for **self pollinated** annual seed production is **5 meters**
58. Isolation distance for **often-cross pollinated** annual seed production is **50 meters**
59. Isolation distance for **cross pollinated** annual seed production is **400-600 meters**
60. Annuals which possess **cytoplasmic male sterility** are **Ageratum, Petunia & Sunflower**
61. Annuals which possess monogenic recessive male sterility are **Calceolaria, Marigold, Salvia, Zinnia**
62. Annuals having maximum number of seed/gram (over 10,000) are **Petunia & Portulaca**
63. Annuals having minimum number of seed/gram (15-20) are **Sweet pea & Sun flower**
64. The first hybrid annual is **Primma donna of '*Begonia semperflorens*'**
65. Primma donna was released by Benary seed company of Germany during **1909**
66. Annual which shows **heterostyly** is **Primula**
67. F_1 annual hybrids developed from Indo-American hybrid seed company, Bangalore are **Petunia, Antirrhinum, Geranium**
68. Annual F_1 hybrids released from NBRI, Lucknow is **Portulaca 'Jhumka' & African Marigold**

69. Annual F_1 hybrid released from IARI, New Delhi is **Holly hock 'Pink beauty' 'Yellow beauty' 'Pastel pink supreme' & 'Apricot supreme'**

70. The first annual F_1 hybrid released in India is from Petunia during **1967**

71. A person who involved in the development of first annual F_1 hybrid from India is **Dr. Manmohan Attavar of Indo- American Hybrid Seed company, Bangalore**

72. First F_1 hybrid in petunia was developed in Japan during **1945**

73. Classical theory of hybrid vigour was given by "Shull" during **1911**

74. Examples for foliage annuals are **Kochia, Amaranthus**

75. Annual seeds which require stratification & scarification is **clianthus**

76. Plant form of Kochia is **Columnar**

77. An annual which shows premature flowering due to **high dose of nitrogen** application is **Lupin**

78. Annuals which shows premature flowering due to **deficiency of nitrogen** is **Clarkia, Candytuft & Salvia**

79. Annuals which possess bold seeds are **holly hock, morning glory, lupin, nasturtium**

80. The season during which annual seeds are being sown in Northern hills is **summer (March- April)**

81. Ideal time for sowing winter annual seeds in south India is **September - October**

82. Late flowering type of annuals are **china aster, dianthus, cineraria**

83. Annuals which produce blue colour flowers are **Corn flower, Blue larkspur, Ageratum, Anchusa, Linaria, Browallia**

84. Annuals which produce yellow & orange flowers are **Marigold, Dimorphotheca, Californian poppy, Nasturtium, Zinnia, Wall flower, Coreopsis, Helichrysum**

85. Annuals which produce white flowers are **Alyssum, China aster, Mathiola, Nigella, Phlox, Papaver, Zinnia, Stock, Dimorphotheca**

86. Annuals which are suitable for neutral or alkaline soils are **Impatiens, Iberis, Papaver, Phlox, Mignonette, Nasturtium & Zinnia**

87. Annuals which responds for pinching are **Ageratum, Calendula, Chrysanthemum, Petunia, Phlox, Dianthus, Schizanthus, Antirrhinum, Verbena & Zinnia**

88. Annuals which perform good in poor soil are **Californian poppy, Cockscomb, Bartonia, Gaillardia, Four-o-clock, Balsam, Godetia, Amaranthus, Nasturtium, Petunia, portulaca, Cleome, Alyssum, Sweet - sultan**

89. Tall annuals are **Holly hock, Chrysanthemum, Sun flower, Helichrysum, Heliotrope, Pimpinella & Zinnia**

90. Dwarf annuals are **Verbena, Viola, Nasturtium, Torenia, Portulaca, Pansy, Phlox, Nemesia, Ice plant, Stock, Linaria, Linum, Lobelia, Candy tuft, Mimulus, Cuphea, California poppy, Gaillardia, Gazania, Godetia, Echium, Wall flower, Ageratum, Sweet Alyssum, Arctotis, Daisy, Brachycome, Acroclinum, Calceolaria, Campanula**

91. Medium-tall annuals are **Amaranthus, Anchusa, Antirrhinum, Calendula, Celosia, Sweet sultan, Corn flower, Clarkia, Cineraria, Coreopsis, Cosmos, Larkspur, Gomphrena, Gypsophila, Carnation, Digitalis, Dimorphotheca, Balsam, Kochia, limonium, Lupin, Bells-of- Ireland, Nicotiana, Nigella, garden poppy, Rudbeckia, Salvia, Schizanthus & African marigold**

92. Annuals which have Indian origin are **Lady's lace, Gomphrena, Tulip, Blue poppy, Primula, Balsam**

93. **Special name or common name** for following annuals are

    Gaillardia- Blanket flower

    Gomphrena- Globe amaranth

    Kochia- Summer cypress or burning bush

    Portulaca- sun plant or rose moss

    Tithonia- Mexican sunflower

    Acroclinum- paper flower or everlasting flower

    Ageratum- Floss flower

    Chrysanthemum coronarium- Garland chrysanthemum, Crown daisy

    Chrysanthemum segetum- Corn marigold

    Chrysanthemum carinatum- summer marguerite

Antirrhinum- snap dragon, dog flower, bunny rabit, bunny mouth

Arctotis- African daisy

Brachycome- Swan River daisy

Calendula- pot marigold

Iberis amara- Hyacinth flower

Clianthus- parrot's Bill

Coreopsis- Tick seed

Corn flower- hurt sickle, blue bottle, bachelor's button, ragged sailor

Bellis perennis- English daisy

Dimorphotheca- African daisy

Gypsophila- Baby's breath

Helichrysum- Everlasting flower or straw flower

Mesembryanthemum spp.-Ice plant- living stone daisy or Fig marigold

Linaria- toad flax

Linum- Flax

Mimulus- Monkey flower

Nigella- love in a mist or david in the bush

Phlox- Star flower

Rudbeckia- cone flower

Salvia- sage flower

Saponaria – soapwart

Schizanthus- butterfly flower or poor man's orchid

Papaver rhoeas - Shirley poppy, corn popy

Statice- sea lavender or pink

Amaranthus caudatus- love-lies-bleeding

Mirabilis jalapa- Four-o-clock plant

Cleome spinosa- spider flower

94. Annuals which belongs to the family **Compositae** are

S. No.	Botanical Name	Common name	Flower colour
Summer and Rainy Season Annuals			
1.	*Gaillardia pulchella*	Blanket flower	Wide range of colours
2.	*Helianthus annus*	Sun Flower	Yellow with brown streaks
3.	*Helianthus cucumerifolius*	Sultan's Autumn beauty	Yellow, orange-yellow, chestnut-brown, maroon
4.	*Tithonia rotundifolia*	Mexican Sunflower	Orange-red with yellow centre
5.	*Zinnia elegans*	Zinnia	White, yellow, violet, pink, red, lilac
	Zinnia angustifolia		-
	Zinnia linearis		Deep yellow, light yellow, white
Winter Season Annuals			
1.	*Acroclinum roseum* *Acroclinum album* *Acroclinum grandiflorum*	Paper flower	Rose colour White Pink
2.	*Ageratum houstonianum*	Floss flower	Blue-mauve
3.	*Arctotis stoechadifolia*	African Daisy	Yellow, orange, white, salmon
4.	*Brachycome iberidifolia*	Swan River Daisy	White, blue, purple, rose-pink, mauve
5.	*Callistephus chinensis*	China Aster	Wide range of colours
6.	*Chrysanthemum coronarium*	Annual chrysanthemum	White, yellow, pink, red, orange, scarlet, mahogany, bronze
7.	*Chrysnthemum segetum*		
8.	*Chrysanthemum carinatum*		
9.	*Senecio cruentus*	Cineraria	White, red, blue, purple, violet
10.	*Coreopsis tinctoria*	Coreopsis	Red, brown with yellow centre
11.	*Centaurea cyanus*	Hurt Sickle/Blue Bottle/Bachelor's Button	Pink, purple, blue, maroon, white

12.	*Cosmos bipinnatus*	Cosmea	Wide range of colours
13.	*Cosmos sulphureus*		Yellow
14.	*Dahlia variabilis*	Dahlia	Wide range of colours
15.	*Dimorphotheca sinuata*	Star of Veldt	Wide range of colours
16.	*Gamolepsis tagetes*	Gamolepsis	Yellow
17.	*Gazania splendens*	Gazania	Red, orange, creamy, yellow, white, brown
18.	*Helichrysum bracteatum*	Everlasting flower/ Straw flower	Red, pink, white, mauve, yellow
19.	*Rudbeckia biocolour*	Cone flower	Yellow with brown markings

95. Annuals which belongs to the family **Amaranthaceae** are

S. No.	Botanical Name	Common name	Flower colour
Summer and Rainy Season Annuals			
1.	*Amaranthus caudatus*	Amaranthus	Purple -red
	Amaranthus hybridus		-
	Amaranthus tricolor ruber		-
	Amaranthus tricolor splendens		-
2.	*Celosia argentia* var. *Cristata*	Cock's Comb	Red, yellow, golden-yellow, orange, white
3.	*Celosia plumosa*		Red, yellow, golden-yellow, orange, silvery white
	Celosia childsi		-
4.	*Gomphrena globosa*	Globe amaranth	White, purple, magenta, pink

96. Annuals which belongs to the family **Cruciferae** are

S.No.	Botanical Name	Common name	Flower colour
Winter Season Annuals			
1.	*Alyssum marittium*	Sweet Alyssum	White, blue
2.	*Iberis amara*	Candytuft	White
3.	*Iberis umbellata*	Hyacinth flower	White, purple
4.	*Matthiola incana*	Stock	Wide range of colours
5.	*Cheiranthes cheiri*	Wall flower	Wide range of colurs

97. Annuals which belongs to the family **Solanaceae** are

S. No.	Botanical Name	Common name	Flower colour
Winter Season Annuals			
1.	*Petunia hybrida*	Petunia	Wide range of colours
2.	*Schizanthus pinnatus*	Butterfly flower of poor man's orchid	Wide range of colours

98. Annuals which belongs to the family **Labiatae** are

S. No.	Botanical Name	Common name	Flower colour
Winter Season Annuals			
1.	*Molucella laevis*	Bells of Ireland	Green, yellow
2.	*Salvia splendens*	Salvia/Sage flower	Wide range of colours

II. Fill in the blanks

1. Isolation distance for self pollinated annual seed production is ....
2. Isolation distance for often-cross pollinated annual seed production is ....
3. Isolation distance for cross pollinated annual seed production is .....
4. The first hybrid annual is.....................
5. The first annual F_1 hybrid released from India is in Petunia during ....
6. Plant form of Kochia is.....................
7. Ideal time for sowing annual seeds in South India is.....................
8. The season during which annual seeds are being sown in Northern hills is ....
9. Annual seed which require stratification & scarification is ....
10. Annual which shows heterostyly is ....?
11. Cool colours are ....
12. Common name of Brachycome is ....
13. Intermediate day annual is ....
14. Long short day annual is.....................
15. Common name of Coreopsis is.....................

III. Choose the best answer

1. Hardy annual is....................

   Viola Oxalis Pansy

2. Tender annual is....................

   Digitalis Rudbeckia Pansy

3. Annual which is suitable for shady situation is....................

   Ice plant Salvia Nasturtium

4. Annual which is suitablr for peculiar shape is....................

   Sweet pea Holly hock Dianthus

5. Foliage annual is....................

   Kochia Cineraria Gaillardia

6. An annual which shows premature flowering due to deficiency of nitrogen is....................

   Holly hock Morning glory Clarkia

7. An annual which shows premature flowering due to high dose of nitrogen application is....................

   Candytuft Salvia Lupine

8. An annual which has Indian origin is....................

   China aster Carnation Gomphrena

9. Annual which responds for pinching is....................

   Zinnia Candytuft Salvia

10. Late flowering type of annual is....................

    Gomphrena Nasturtium China aster

11. Annual which is suitable for rock garden is....................

    Carnation Ice plant Salvia

12. An annual which produce white flower is....................?

    Nigella Coreopsis Helichrysum

13. Long day annual is....................

    Chrysanthemum Petunia Cosmos

14. Annual which possess self- incompatibility is....................

    Ageratum Marigold Chrysanthemum

15. Annual which is having maximum number of seed/gram (over 10,000) is.....................

    Portulaca Zinnia Marigold

IV. Match the following

1.	Tall annual	:	Benary seed company of Germany
2.	Impatiens	:	Ageratum
3.	Hybrid vigour theory	:	Corn flower
4.	Primma donna	:	Holly hock
5.	Self- incompatibility	:	Alkaline soil
6.	Blue colour		1911
7.	Self pollinated annual	:	Sweet sultan
8.	Screening purpose	:	Yellow colour
9.	Cut flower	:	Nasturtium
10.	Fragrant flower	:	Lupin
11.	Coreopsis	:	Holly hock
12.	Bold seed	:	Antirrhinum
13.	English daisy	:	Brachycome
14.	African daisy	:	*Bellis perennis*
15.	Swan River daisy	:	Dimorphotheca

Answers

II. Fill in the blanks

1. 5 meters
2. 50 meters
3. 400-600 meters
4. Primma donna 'Begonia semper florens'
5. 1967
6. Columnar
7. September – October
8. Summer (March- April)
9. Clianthus
10. Primula
11. Green & blue
12. Swan River daisy
13. Coleus
14. Aloe
15. Tick seed

III. Choose the best answer

1. Viola
2. Pansy
3. Salvia
4. Dianthus
5. Kochia
6. Clarkia
7. Lupine
8. Gomphrena
9. Zinnia
10. China aster
11. Ice plant
12. Nigella
13. Petunia
14. Ageratum
15. Portulaca

IV. Match the following

1. Holly hock
2. Alkaline soil
3. 1911
4. Benary seed company of Germany
5. Ageratum
6. Corn flower
7. Lupin
8. Holly hock
9. Antirrhinum
10. Sweet sultan
11. Yellow colour
12. Nasturtium
13. *Bellis perennis*
14. Brachycome
15. Dimorphotheca

ANNEXURE- IV

ANNUALS

S. No.	Botanical Name	Common name	Family	Flower colour	Remarks
Summer and Rainy Season Annuals					
1.	*Amaranthus caudatus*	Amaranthus	Amarantaceae	Purple -red	Drooping tail like inflorescence, very good for drying
	Amaranthus hybridus			-	-
	Amaranthus tricolor rubber			-	-
	Amaranthus tricolor splendens			-	-
2.	*Celosia argentia* var. *Cristata*	Cock's Comb	Amarantaceae	Red, yellow, golden-yellow, orange, white	Large crested flower heads like comb, good borders, beds and dry flowers
3.	*Celosia plumosa*			Red, yellow, golden-yellow, orange, silvery white	Long feathery spikes good borders, beds and dry flowers
	Celosia childsi			-	-
4.	*Gaillardia pulchella*	Blanket flower	Compositeae	Wide range of colours	Good for beds, borders, edge, rockery, pot and as cut flowers

5.	*Gomphrena globosa*	Globe amaranth	Amarantaceae	White, purple, magenta, pink	-
6.	*Impatiens balsamina*	Balsam	Balsaminaceae	Wide range of colours	Good for semi-shady, sunny areas for pots, beds, borders, rockery in drained soil
7.	*Helianthus annus*	Sun Flower	Compositae	Yellow with brown streaks	Good for borders, beds, background or screen. Direct seeding is common
8.	*Helianthus cucumerifolius*	Sultan's Autumn beauty		Yellow, orange-yellow, chestnut-brown, maroon	Well branched with small flowers on long stems
9.	*Kochia scoparia* var. *tricophylla*	Summer cypress/ Burning bush	Chenopodiaceae	-	Columnar to pyramidal shape, very good light green foliage. Good for pots, beds and borders
10.	*Portulaca grandiflora*	Sun plant/Rose moss	Portulaceae	Wide range of colours	Annual with drooping succulent branches, good for shady and semi-shady areas in pots
11.	*Tithonia rotundifolia*	Mexican Sunflower	Compositae	Orange-red with yellow centre	Flowers like dahlia 7-10cm across, good for beds, borders and background
12.	*Zinnia elegans*	Zinnia	Compositae	White, yellow, violet, pink, red, lilac	Flower size large to small. Good for beds, borders, background and as loose

					flowers. Direct seeding is also done.
	Zinnia angustifolia			-	-
	Zinnia linearis			Deep yellow, light yellow, white	Suitable for edging, beds and mass effect
Winter Season Annuals					
13.	*Acroclinum roseum*	Paper flower	Compositae	Rose colour	Flowers are single or double and keep their form and colour on drying
	Acroclinum album			White	
	Acroclinum grandiflorum				Pink
14.	*Ageratum houstonianum*	Floss flower	Compositae	Blue-mauve	Good for bedding, as cut flower and dry flower
15.	*Anchusa capensis*	Anchusa	Boraginaceae	Yellow	Very good for hills with coarse texture foliage
16.	*Antirrhinum majus*	Snap dragon/ Dog flower/Bunny habit/Bunny mouth	Scrophulariaceae	Wide range of colours	Very good for bedding, cut flowers and pots
17.	*Arctotis stoechadifolia*	African Daisy	Compositae	Yellow, orange, white, salmon	Good for bedding and cut flowers
18.	*Brachycome iberidifolia*	Swan River Daisy	Compositae	White, blue, purple, rose-pink, mauve	Fine texture foliage, very good for pots, edge and hanging baskets
19.	*Callistephus chinensis*	China Aster	Compositae	Wide range of colours	Single, semi-double. Double flowers

20.	*Calendula officinalis*	Pot marigold	Asteraceae	Yellow, golden yellow	Single, semi-double, double flowers, good for pots, beds, borders and very good cut flowers
21.	*Chrysanthemum coronarium*	Annual chrysanthemum	Compositae	White, yellow, pink, red, orange, scarlet, mahogany, bronze	Flowers are single, semi-double and double, good for bedding and cut flowers
22.	*Chrysnthemum segetum*				
23.	*Chrysanthemum carinatum*				
24.	*Eschscholtzia californica*	California poppy	Papaveraceae	Wide range of colours	Good for beds, borders, edge, rockery and pots. Direct seeding is also done.
25.	*Molucella laevis*	Bells of Ireland	Labiatae	Green, yellow	Valued for green/yellow calyces. Good for cut, dry flowers and for growing in pots, beds, borders
26.	*Iberis amara*	Candytuft	Cruciferae	White	Good for edge, pots, beds and borders in sunny areas
27.	*Iberis umbellata*	Hyacinth flower		White, purple	
28.	*Senecio cruentus*	Cineraria	Compositae	White, red, blue, purple, violet	Very good for pots and beds in shady areas
29.	*Clarkia elegans*	Clarika	Onagraceae	Wide range of colours	Single and double flowers, good for borders, beds and as

					cut flowers in partial shady areas
30.	*Clianthusdampieri*	Parrot's Bill	Legumonosae	Dark red, scarlet	Very good pot plant. Seeds require stratification
31.	*Coreopsis tinctoria*	Coreopsis	Compositae	Red, brown with yellow centre	Good as pot plant, bedding or borders in sunny situation. Require heavy watering
32.	*Centaurea cyanus*	Hurt Sickle/Blue Bottle/ Bachelor's Button	Compositae	Pink, purple, blue, maroon, white	Good for borders, beds and as cut flowers
33.	*Cosmos bipinnatus*	Cosmea	Compositae	Wide range of colours	Fine texture foliage, good for borders, bedding and background
34.	*Cosmos sulphureus*			Yellow	Flowers borne on long stems, good for bedding, borders and background
35.	*Dahlia variabilis*	Dahlia	Compositae	Wide range of colours	-
36.	*Bellisperennis*	English Daisy	Loasaceae	White, pink	Good for bedding, pots, hanging baskets and drying
37.	*Dimorphotheca sinuata*	Star of Veldt	Compositae	Wide range of colours	Good for beds, borders, edge, rockery and pots
38.	*Gamolepsis tagetes*	Gamolepsis	Compositae	Yellow	Highly suitable for bed, edging pots and rock garden

39.	*Gazania splendens*	Gazania	Compositae	Red, orange, creamy, yellow, white, brown	Very good for pots, beds and as cut flowers
40.	*Gypsophila elegans*	Baby breath	Caryophyllaceae	White	Very good cut flowers used as filler in arrangements and for pots, beds, borders
41.	*Helichrysum bracteatum*	Everlasting flower/ Straw flower	Compositae	Red, pink, white, mauve, yellow	Very good for beds, borders, background, cut flowers and dry flowers
42.	*Althaea rosea*	Holly Hock	Malvaceae	All colours	Herbaceous border, direct seeding is common
43.	*Mesembryanthemum criniflorum*	Ice Plant/Living stone Daisy	Aizoaceae	Wide range of colours	Very good for pots, edge, hanging baskets, beds in sunny dry areas. Seeds are also directly sown.
44.	*Pimpinella monoica*	Lady's lace	Umberlliferae	White	Good for beds, borders, background in sunny areas
45.	*Delphinium hybridum*	Larkspur	Ranunculaceae	Wide range of colours	Fine texture dark green foliage. Good for beds, borders, pots and as cut flowers
46.	*Linaria maroccana*	Linaria/Toad Flax	Scrophulariaceae	Wide range of colours	Suitable for pots, edge, beds, borders and cut flowers. Seeds are also directly sown.

47.	*Linum grandiflorum*	Linum/Flax	Linaceae	Red, pink, blue, white	Suitable for pot, edge, beds, borders and cut flowers. Seeds are also directly sown.
48.	*Lupinus hartwegii*	Lupine	Papilionaceae	Wide range of colours	Very good for hills in pots, beds, borders and cut flowers. Seeds are directly sown.
49.	*Reseda ordorata*	Mignonette	Resedaceae	Wide range of colours	Good for pots, beds, window-sills in semi-shady areas and as cut flowers
50.	*Mimulus luteus*	Monkey flower	Scrophulariaceae	Yellow, crimson, pink, red, bicoloured	Good for growing in beds, borders, pots in moist soil in semi-shady areas
51.	*Tropaeleum majus*	Nasturtium	Tropaelaceae	Orange, yellow	Trailing plant with single and semi-double flowers, good for screening, hanging baskets in semi-shady and sunny areas
52.	*Nemesia strumosa*	Nemasia	Scrophulariaceae	Wide range of colours	Very good as cut flowers and for growing in pots, beds, borders, rockery
53.	*Nigella domascena*	Love in a-mist/ Devil-in-the bush	Ranunculaceae	White, pink	Flowers 2.5-5cm across, tubular. Good for growing in beds, pots and as a cut flowers
54.	*Viola wittorckiana*	Pansy	Violaceae	Wide range of colours and spotted	Flowers look like butterfly, good for edge, pots windows-sills and hanging baskets

55.	*Petunia hybrida*	Petunia	Solanaceae	Wide range of colours	Flowers are bell shaped single, semi-double and double. Good for pots, beds, borders
56.	*Phlox drummondii*	Phlox/Star flower	Polemoniaceae	Wide range of colours	Good for edge, beds, pots, window-sills, rockery in sunny and semi-shady areas
57.	*Rudbeckia biocolour*	Cone flower	Compositae	Yellow with brown markings	Good for growing in pots, beds, borders and as cut flowers in hills
58.	*Salvia splendens*	Salvia/Sage flower	Labiatae	Wide range of colours	Good for pots, beds, borders, background and window-sills in semi-shady areas and as cut flowers
59.	*Saponaria calabrica*	Saponaria/Soapwart	Caryophyllaceae	Pink, white	Good for edge, pots, hanging baskets and rockery in sunny areas
60.	*Schizanthus pinnatus*	Butterfly flower of poor man's orchid	Solanaceae	Wide range of colours	Very good as cut flowers and pot, beds, borders and background in hills
61.	*Papaver rhoeas*	Corn poppy	Papaveraceae	Wide range of colours	Flowers cup shaped single, double. Good for pots, beds, borders and good cut flowers. Seeds are also directly sown.
62.	*Limonium sinuatum*	Statice/Sea lavendar	Plumbaginaceae	Wide range of colours	Good for sunny areas in beds, borders, pots and very good as cut and dry flowers

63.	*Matthiola incana*	Stock	Cruciferae	Wide range of colours	Very good cut flowers in single or double. Also good for growing in pots, beds, borders
64.	*Alyssum marittium*	Sweet Alyssum	Cruciferae	White, blue	Sweet smelling flowers, good for edge, bedding, pots and hanging baskets
65.	*Lathyrus odoratus*	Sweet pea	Leguminosae	Wide range of colours	Climbing fragrant annual good for screening, borders and beds. Direct seeding is done. Seeds of lavender, blue, mauve colour varieties are wrinkled
66.	*Centaurea moschata*	Sweet Sultan	Compositae	White, red, mauve, violet, purple, yellow	Flowers borne on long stems good for bedding and cut flowers
67.	*Dianthus barbatus*	Sweet William	Caryophyllaceae	Pink, red, white, purple, scarlet, maroon	Sweet smelling flowers. Good for beds, edge, pots and cut flowers
68.	*Verbena hybrida*	Verbena	Verbenaceae	Wide range of colours	Good for dry sunny areas in pots, beds, borders and rockery
69.	*Cheiranthes cheiri*	Wall flower	Cruciferae	Wide range of colurs	Good for bedding and cut flowers

Chapter **8**

Palms & Cycads

I. General information

The nature of roots

The roots of palms are usually tough and thick, although slender and wiry types are existing. Some roots creep over the soil's surface, while a few palms even develop aerial roots that may eventually make them appear to be growing on stilts.

A few palms develop aerial roots at or near the base of the trunk or stems. These include *Crysophila warscewiczii* (Rootspine palm) and *Clinostigma exorrhizum.* Other palms with aerial roots include *Pinanga aristata, Verschaffeltia splendida* (Seychelles stilt palm) and *Socratea exorrhiza.* Some of these develop into stilt or prop roots, which often form a cone and give the palm extra support, they usually occurs on plams native to wet soils and humid conditions, whereas the base of the palms is at risk from decay. Indeed, in a few instances, the lower part of the trunk withers and the palm is entirely supported by these roots.

The nature of trunks

Trunks vary in diameter, most achieving their final diameter as the trunk grows, seldom does it widen as a result of later upward growth of the palm. Most trunks are straight, but a few are swollen, perhaps the best known of these being *Hyophorbe lagenicaulis* (Bottle plam) and *Hyophorbe verschaffeltii* (Spindle palm). The trunk of *Hyophorbe lagenicaulis* has a bulbous base, while that of *Hyophorbe verschaffeltii* is narrow at its crown and base, but widens at its centre.

Trunk surface also varies, whereas some are relatively smooth, such as *Jubaea chilensis* (Chilean wine palm, Coquito palm), the trunk of *Aiphanes aculeata* (Ruffle palm, Chonta Ruro) is ringed in black spines. Some palms have a mass of fibres on their trunks, such as *Trachycarpus fortunei* (Chinese windmill palm, Chusan palm).

The range of trunks includes some with distinctive growth rings (formed of scars from old leaves) such as *Howea belmoreana* (Belmore sentry palm, Curly palm), *Adonidia merrillii* (Christmas palm, Manila palm) has indistinct growth rings, while *Phoenix roebelenii* (Dwarf date plam, Roebelin palm) has dominant, persistent leaf bases.

The nature of leaves

Also known as fronds, the leaves of palms have varied natures, some are borne at the top of a palm in a distinctive crown, while others have leaves on stems that develop from lower down on the trunk or even from stems that arise directly from the ground.

Fan-leaved palms

Also known as Fan palms, these have leaves with circular or semi-circular outline. The leaves resemble fans and are either partly divided known as palmafid or referred to as palmate if totally divided. The leaf divisions are known as segments.

Palms with fan shaped leaves include *Chamaerops humilis* (European Fan Palm, Mediterranean Fan Palm), *Pritchadia pacifica* (Fiji Fan Palm), *Trachycarpus fortune* (Chinese Windmill Palm, Chusan Palm) and *Washingtonia filifera* (Californian Fan Palm, Petticoat palm). *Licuala spinosa* (Mangrove Fan Palm, Spicy Licuala) has wedge shaped leaflets.

Feather-type palms

Also known as Pinnate palms, these palms have fronds that are divided on either side of the mid-rib, with the impression of many leaflets. They are sometimes described as the backbones and ribs of a fish.

Palms that reveal this style include *Howea belmoreana* (Belmore Sentry Palm or Curly Palm), *Howea forsteriana* (Kentia Palm or Thatch Leaf Palm) and *Lytocaryum weddelianum* (Dwarf Coconut Palm or Weddel Palm).

The leaflets also known as pinnae, are variable in size, shape and the angle at which they are attached to the central leaf-stem or rachis.

Some arise from the rachis in a flat, even distribution, such as *Archontophoenix alexandrae* (Alex Palm, Northern Bangalow Palm) and *Archontophoenix cunninghamiana* (Piccabeen Bangalow Palm). Some, such as *Howea belmoreana* (Belmore Sentry Palm or Curly Palm) create an angled, upright, perhaps V-shaped formation, while others droop. Some have leaflets in two distinct planes, such as *Phoenix canariensis* (Canary Island Date Palm), while others have an uneven distribution along the rachis, including the beautiful and popular Hydriastele *wendlandiana* (Florence Falls Palm, Latrum Palm).

A further type has a plumose (feather-like) and irregular arrangement of leaflets, such as *Syagrus romanzoffiana* (Giriba Palm, Queen Palm).

Fishtail palms

Also known as bipinnate palms, these have leaves that are twice divided and have a fishtail appearance at their tips. Examples are *Caryota mitis* (Clustered Fishtail Palm, Tufted Fishtail Palm) and *Caryota urens* (Fishtail Palm, Toddy palm).

Foxtail palm

It is thin trunked, pinnate palm which gives very attractive apperaence. Leaves are very plumose with leaflets coming off the leaf stem at all angles. E.g. *Wodyetia bifurcate.*

Entire-leaved palms

These attractive palms are unusal in that leaves are undivided in their mature state. One example of this type is *Johannesteijsmannia altifrons* (Diamond Joey).

The nature of growth

Palms have a wide range of growth habits, from those that are trunkless to those with a clumping nature or majestic palms with tall trunks. Some even have a climbing nature.

Trunkless palms

Several palms appear to be trunkless, including those that later form dominant trunks but, while young, seem to have only a mass of leafstalks growing from low down on the palm and from the ground level, Often, palms native to scrubby land are trunkless.

A few palms have trunks that are wholly or partly beneath the soil's surface, including *Sabal minor* (Blue palmetto palm, Little blue stem) and *Serenoa repens* (Saw Palmetto, Scrub Palmetto).

Clumping palms

These palms have multiple trunks or stems. They include popular species such as *Dypsis lutescens* (Golden cane palm, Yellow bamboo palm), *Phoenix dactylifera* (Date palm, Date) and *Phoenix reclinata* (African date palm, Senegal date palm).

Sometimes, a clump of palms is initially formed of many stems, but later one or two dominant and much larger forming trunks are retained (*Chamaerops humilis*).

Solitary palms

Palms with solitary trunks include *Adonidia merrillii* (Christmas palm, Manila palm), *Livistona australis* (Australian cabbage palm, Gippsland palm), *Phoenicophorium borsigianum* (Latanier palm), *Trachycarpus fortunei* (Chinese windmill palm, Chusan palm) and *Washingtonia filifera* (Californian fan palm, Petticoat palm).

Branching palms

Few palms have a natural branching nature although branching can sometimes occur as the result of damage to the growing point of a trunk. *Hyphaena thebaica* (Gingerbread palm) is native to northern and northeast Africa, where it has a natural, freely branching and clustering nature. *Serenoa repens* (Saw Palmetto, Scrub Palmetto) is a trunkless palm with branching underground stems that send up new stems.

Climbing Palms

More than 600 palms have a climbing nature, including *Calamus australis* (Lawyer cane, Rattan palm), with its scrambling, climbing, clustering nature and stems that are up to 25 m high in the wild area.

Many palms with a climbing nature initially form clumps, with stems usually known as canes, a few have a single climbing stem.

Occasionally, the canes spread over the ground until they find a plant to which they can cling. Many climbing palms have canes clothed in spines and hooks. These enable to cling to a host and give it protection from animals. The clusters of flowers also aid in supporting canes, as many have spine-like hooks.

The nature of flowers

The arrangement of flowers in palms varies widely, including those at the top of the stems that grow from ground level and those in the heads or crowns of palms.

Palms with a relatively low stature, flower earlier than tall types. For example *Chamaedorea elegans* (Good luck plam, Parlour palm) which grows up to 3 m high and is often grows as an indoor palm, may flower when four to six years old, whereas *Lodoicea maldivica,* which can be 25 m or more in the wild areas, will flower only at the age of 40 years or more.

Some palms are monocarpic, indicating that a stem or trunk dies once it has borne flowers. Unless the palm has a clumping nature or is formed of many individual stems , it results in death of entire plant (E.g. *Corypha umbraculifera).*

Several palms produce their flowers terminally and towards the top of the trunk, whereas others develop axillary inflorescences from shoots at the junctions of leaf stalks or just below the crownshaft of palms with pinnate (feather-like) leaves.

The individual flowers of palms usually small and are borne in clusters that sometimes bear more than a million individual flowers. Some are attractively fragrant, such as *Coccothrinax fragrans* (Fragrant Cuban thatch, Yuraguana) and *Hyophorbe verschaffelti* (Spindle palm) while others like *Arenga pinnata* (Sugar palm, Gomuti palm) have drooping clusters of purple flowers with a rather unpleasant aroma.

Some palms have flowers that have a bisexual nature, with both male and female parts present. They are called hermaphrodite. Others have a unisexual nature, with only one sex present. Unisexual flowers can either be borne separately on the same inflorescence or in separate inflorescences on the same palm. Many species have male and female flowers on separate plants.

Unlike many flowers on trees, those on palms are relatively short-lived and usually last for only about a day. They are either pollinated by insects, such as honeybees, flies and beetles or by the wind.

The nature of fruits

Palm fruits and seeds vary in size and colour. They are botanically either a drupe (fleshy and usually with a single stone-like seed, like a plum or cherry) or single-seeded berries. Many palm fruits are edible and succulent, perhaps the best known being *Phoenix dactylifera* (Date palm), which has been part of the diet of millions of people for centuries, especially in warm regions.

List of palms

1. *Acoelorraphe wrightii* (Everglades palm, Paurotis palm, Saw cabbage palm, Silver saw palm, Silver saw palmetto)
2. *Adonidia merrillii* (Adonidia palm, Christmas palm, Dwarf royal palm, Manila palm)
3. *Aiphanes aculeata* (Chonta Ruro, Coyure Palm, Ruffle Palm, Spine Palm)
4. *Aiphanes erosa* (Macaw palm)
5. *Archontophoenix alexandrae* (Alex palm, Alexandra king palm, Alexander king palm, King palm, Northern bungalow palm)
6. *Archontophoenix cunninghamiana* (Bangalow palm, Piccabeen bungalow palm, Piccabean palm)
7. *Areca catechu* (Arecanut palm, Betel nut palm, Betel palm, Caccu, Catechu, Pinang)
8. *Arenga caudata* (Dwarf sugar palm)
9. *Arenga engleri* (Dwarf sugar palm, Sugar palm, Formosa palm)
10. *Arenga pinnata* (Aren, Areng palm, Black fibre palm, Gomuti palm, Kabong, Sugar palm)
11. *Arenga undulatifolia* (Aren Gelora)
12. *Bactris gasipaes* (Chonta, Peach palm, Pejibaya, Pejibeye, Pejivalle, Pewa, Pupunha)
13. *Bismarckia nobilis* (Bismarck palm)

14. *Borassus flabellifer* (Doub palm, Lontar plam, Palmyra palm, Pannamaram, Tala palm, Talauriksha palm, Tal-gas, Toddy palm, Wine palm)
15. *Brahea armata* (Blue fan palm, Blue hesper palm, Grey goddess, Mexican blue fan palm, Mexican blue palm, Short blue hesper)
16. *Brahea edulis* (Guadaloupe palm, Guadalupe palm)
17. *Butia capitata* (Butia palm, Jelly palm, Pindo palm, South American jelly palm, Wine palm)
18. *Calamus australis* (Hairy mary, Lawyer cane, Lawyer's cane, Rattan palm, Wait-a-while vine)
19. *Carpentaria acuminata* (Carpentaria palm)
20. *Caryota mitis* (Burmese fishtail palm, Clustered fishtail palm, Fishtil palm and Tufted fishtail palm)
21. *Caryota urens* (Fishtail palm, Jaggery palm, Kitul tree, Sago palm, Solitary fishtail palm, Toddy palm, Wine palm)
22. *Ceroxylon quindiuense* (Andean wax palm, Wax palm)
23. *Chamaedorea cataractarum* (Cascade palm, Cat palm, Cataract palm)
24. *Chamaedorea elegans* (Good luck palm, Parlor palm, Parlour palm)
25. *Chamaedorea erumpens* (Bamboo palm)
26. *Chamaedorea metallica* (Metallic palm, Miniature fishtail plam)
27. *Chamaedorea microspadix* (Bamboo palm, Hardy bamboo palm)
28. *Chamaedorea seifrizii* (Bamboo palm, Reed palm)
29. *Chamaedorea stolonifera*
30. *Chamaerops humilis* (European fan palm, Fan palm, Mediterranean fan palm)
31. *Coccothrinax argentata* (Florida silver palm, Silver palm, Silver thatch palm, Silvertop, Silver top palm)
32. *Coccothrinax crinita* (Mat palm, Old man palm, Old man thatch palm, Palma petate, Thatch palm)
33. *Coccothrinax spissa* (Guano, Swollen silver thatch)
34. *Cocos nucifera* (Coco palm, Coconut palm, Coconut)
35. *Copernicia baileyana* (Bailey fan palm, Bailey's copernicia palm, Yarey, Yarey Hembra, Yareyon)

36. *Copernicia macroglossa* (Cuban petticoat palm, Jata de Guanbacoa, Petticoat palm)
37. *Corypha umbraculifera* (Talipot palm)
38. *Cyrtostachys renda* (Sealing wax palm, Lipstick palm, Maharajah palm, Pinang-rajah)
39. *Dictyosperma album* (Common princess palm, Hurricane palm, Princess palm)
40. *Dypsis decaryi* (Three-sided palm, Triangle palm)
41. *Dypsis lutescens* (Areca palm, Butterfly palm, Cane palm, Golden feather palm, Golden yellow palm, Madagascar palm, Yellow palm, Yellow butterfly palm)
42. *Elaeis guineensis* (African oil palm, Jacara palm)
43. *Euterpe edulis* (Assai palm, Jacara palm)
44. *Hedyscepe canterburyana* (Big mountain palm, Umbrella palm)
45. *Howea belmoreana* (Belmore sentry palm, Curly palm, Sentry palm)
46. *Howea forsteriana* (Forster sentry palm, Kentia palm, Paradise palm, Sentry palm, Thatch leaf palm)
47. *Hydriastele wendlandiana* (Florence falls palm, Latrum palm)
48. *Hyophorbe lagenicaulis* (Bottle palm)
49. *Hyophorbe verschaffeltii* (Spindle palm)
50. *Johannesteijsmannia altifrons* (Diamond Joey)
51. *Jubaea chilensis* (Chinese wine palm, Honey palm, Honey wine palm, Little cokernut, Syrup palm, Wine palm)
52. *Jubaeopsis caffra* (Pondoland palm)
53. *Laccospadix australasica* (Atherton palm, Queensland kentia)
54. *Latania loddigesii* (Blue latan, Blue latan palm)
55. *Latania lontaroides* (Red latan, Red latan palm)
56. *Latania verschaffeltii* (Yellow, Yellow latan palm)
57. *Licuala grandis* (Fan-leaved palm, Palas payung, Ruffled fan palm, Vanuatu fan palm)
58. *Licuala orbicularis*

59. *Licuala spinosa* (Mangrove fan palm, Spiny licuala)
60. *Linospadix monostachya* (Walking stick palm)
61. *Livistona australis* (Australian cabbage palm, Australian palm, Australian fan palm, Cabbage palm, Fan palm and Gippsland palm)
62. *Livistona chinensis* (Chinese fan palm, Chinese fountain palm, Fan palm, Footstool palm, Fountain palm)
63. *Lodoicea maldivica* (Coco-de-Mer, Double coconut, Seychellus nut)
64. *Lytocaryum weddellianum* (Dwarf coconut palm, Sago palm, Weddel palm)
65. *Metroxylon sagu* (Sago palm)
66. *Normanbya normanbyi* (Black palm)
67. *Oncopserma tigillarium* (Nibung palm)
68. *Parajubaea cocoides* (Coco Cumbe, Coquito, Mountain coconut, Quito palm)
69. *Phoenicophorium borsigianum* (Latanier palm)
70. *Phoenix canariensis* (Canary island date, Canary island date palm, Canary date palm)
71. *Phoenix dactylifera* (Date, Date palm)
72. *Phoenix reclinata* (African date palm, African wild date palm, Senegal date palm)
73. *Phoenix roebelenii* (Dwarf date palm, Miniature date palm, Pygmy date palm, Roebelin palm)
74. *Phoenix rupicola* (Cliff date palm, Date palm, East India wine palm, India date palm, Wild date palm)
75. *Phoenix sylvestris* (India Date, Khajuri, Silver date palm, Sugar date palm, Sugar palm, Sugar palm of India, Toddy palm, Wild date, Wild date palm)
76. *Pinanga coronata* (Bunga, Ivory cane palm, Pinang palm)
77. *Pritchardia pacifica* (Fan palm, Fiji fan palm, Pacific fan palm)
78. *Pseudophoenix sargentii* (Buccaneer palm, Cherry palm, Florida cherry palm)
79. *Ptychosperma elegans* (Alexander palm, Solitaire palm)

80. *Ptychosperma macarthurii* (Hurricane palm, Macarthur feather palm, Macarthur palm)
81. *Raphia farinifera* (Madagascar raffia palm, Latrum palm)
82. *Ravenea rivularis* (Majesty palm)
83. *Reinhardtia gracilis* (Window palm, Window pane palm)
84. *Rhapidophyllum hystrix* (Blue palmetto, Creeping palmetto, Dwarf saw palmetto, Hedgehog palm, Needle palm, Porcupine palm, Spine palm, Vegetable porcupine)
85. *Rhapis excelsa* (Bamboo palm, Fern rhapis, Ground rattan, Lady palm, Little lady palm, Slender lady palm, Miniature fan palm)
86. *Rhapis humilis* (Reed rhapis, Slender lady palm)
87. *Rhopalostylis sapida* (Brush palm, Feather duster palm, Nikau, Nikau palm, Shaving-brush palm)
88. *Roystonea regia* (Cuban royal , Cuban royal palm, Florida royal palm, Royal palm)
89. *Sabal minor* (Blue palmetto palm, Bush palmetto, Dwarf palm, Dwarf palmetto, Dwarf palmetto palm, Little blue stem, Scrub palmetto, Swamp palmetto)
90. *Sabal palmetto* (Blue palmetto, Cabbage palm, Cabbage palmetto, Cabbage tree, Common Palmetto, Palmetto palm)
91. *Serenoa repens* (Saw palmetto, Scrub palmetto)
92. *Syagrus romanzoffiana* (Giriba palm, Queen palm)
93. *Thrinax morrisii* (Brittle thatch, Brittle thatch palm, Broom palm, Buffalo-thatch, Buffalo top, Peaberry palm, Pimetta, Key palm, Key thatch, Silver thatch palm)
94. *Thrinax parviflora* (Broom palm, Florida thatch palm, Iron thatch, Mountain thatch palm, Jamaica thatch palm, Palmetto thatch, Thatch, Thatch pole)
95. *Thrinax radiata* (Florida thatch palm, Jamaica thatch, Sea thatch, Silk-top thatch)
96. *Trachycarpus fortunei* (Chinese windmill palm, Chusan palm, Fan palm, Hemp palm, Windmill palm)
97. *Trithrinax acanthacoma* (Buriti palm, Spiny fibre palm)
98. *Washingtonia filifera* (American cotton palm, Caliornian cotton palm, Californian fan palm, Caliornian palm, Cotton palm, Desert palm,

Desert fan palm, Desert Washingtonia, Fan palm, Northern Washingtonia, Petticoat palm)

99. *Washingtonia robusta* (Mexican Washington palm, Mexican Washingtonia, Mexican fan palm, Skyduster Southern Washingtonia, Thread palm)

100. *Wodyetia bifurcata* (Foxtail palm)

Palms suitable as houseplants

1. *Aiphanes aculeata* (Chonta Ruro, Coyure Palm, Ruffle Palm, Spine Palm)
2. *Archontophoenix alexandrae* (Alex palm, Alexandra king palm, Alexander king palm, King palm, Northern bungalow palm)
3. *Archontophoenix cunninghamiana* (Bangalow palm, Piccabeen bungalow palm, Piccabean palm)
4. *Arenga caudata* (Dwarf sugar palm)
5. *Carpentaria acuminata* (Carpentaria palm)
6. *Caryota mitis* (Burmese fishtail palm, Clustered fishtail palm, Fishtil palm and Tufted fishtail palm)
7. *Caryota urens* (Fishtail palm, Jaggery palm, Kitul tree, Sago palm, Solitary fishtail palm, Toddy palm, Wine palm)
8. *Chamaedorea cataractarum* (Cascade palm, Cat palm, Cataract palm)
9. *Chamaedorea elegans* (Good luck palm, Parlor palm, Parlour palm)
10. *Chamaedorea erumpens* (Bamboo palm)
11. *Chamaedorea metallica* (Metallic palm, Miniature fishtail plam)
12. *Chamaedorea seifrizii* (Bamboo palm, Reed palm)
13. Chamaedorea stolonifera
14. *Chamaerops humilis* (European fan palm, Fan palm, Mediterranean fan palm)
15. *Dypsis decaryi* (Three-sided palm, Triangle palm)
16. *Dypsis lutescens* (Areca palm, Butterfly palm, Cane palm, Golden feather palm, Golden yellow palm, Madagascar palm, Yellow palm, Yellow butterfly palm)
17. *Hedyscepe canterburyana* (Big mountain palm, Umbrella palm)

18. *Howea belmoreana* (Belmore sentry palm, Curly palm, Sentry palm)
19. *Howea forsteriana* (Forster sentry palm, Kentia palm, Paradise palm, Sentry palm, Thatch leaf palm)
20. *Laccospadix australasica* (Atherton palm, Queensland kentia)
21. *Licuala grandis* (Fan-leaved palm, Palas payung, Ruffled fan palm, Vanuatu fan palm)
22. Licuala orbicularis
23. *Linospadix monostachya* (Walking stick palm)
24. *Livistona australis* (Australian cabbage palm, Australian palm, Australian fan palm, Cabbage palm, Fan palm and Gippsland palm)
25. *Livistona chinensis* (Chinese fan palm, Chinese fountain palm, Fan palm, Footstool palm, Fountain palm)
26. *Phoenix canariensis* (Canary island date, Canary island date palm, Canary date palm)
27. *Phoenix roebelenii* (Dwarf date palm, Miniature date palm, Pygmy date palm, Roebelin palm)
28. *Pinanga coronata* (Bunga, Ivory cane palm, Pinang palm)
29. *Ptychosperma elegans* (Alexander palm, Solitaire palm)
30. *Reinhardtia gracilis* (Window palm, Window pane palm)
31. *Rhapis excelsa* (Bamboo palm, Fern rhapis, Ground rattan, Lady palm, Little lady palm, Slender lady palm, Miniature fan palm)
32. *Rhapis humilis* (Reed rhapis, Slender lady palm)
33. *Syagrus romanzoffiana* (Giriba palm, Queen palm)
34. *Trachycarpus fortunei* (Chinese windmill palm, Chusan palm, Fan palm, Hemp palm, Windmill palm)
35. *Washingtonia filifera* (American cotton palm, Caliornian cotton palm, Californian fan palm, Caliornian palm, Cotton palm, Desert palm, Desert fan palm, Desert Washingtonia, Fan palm, Northern Washingtonia, Petticoat palm)
36. *Washingtonia robusta* (Mexican Washington palm, Mexican Washingtonia, Mexican fan palm, Skyduster Southern Washingtonia, Thread palm)

Palms for only tropical regions

1. *Cyrtostachys renda* (Sealing wax palm, Lipstick palm, Maharajah palm, Pinang-rajah)
2. *Licuala grandis* (Fan-leaved palm, Palas payung, Ruffled fan palm, Vanuatu fan palm)

Palms for both tropical and subtropical regions

1. *Aiphanes aculeata* (Chonta Ruro, Coyure Palm, Ruffle Palm, Spine Palm)
2. *Archontophoenix alexandrae* (Alex palm, Alexandra king palm, King palm, Northern bungalow palm)
3. *Caryota mitis* (Burmese fishtail palm, Clustered fishtail palm, Fishtil palm and Tufted fishtail palm)
4. *Caryota urens* (Fishtail palm, Jaggery palm, Kitul tree, Sago palm, Solitary fishtail palm, Toddy palm, Wine palm)
5. *Chamaedorea seifrizii* (Bamboo palm, Reed palm)
6. *Dictyosperma album* (Common princess palm, Hurricane palm, Princess palm)
7. *Hydriastele wendlandiana* (Florence falls palm, Latrum palm)
8. *Hyophorbe lagenicaulis* (Bottle palm)
9. *Hyophorbe verschaffeltii* (Spindle palm)
10. *Laccospadix australasica* (Atherton palm, Queensland kentia)
11. Licuala orbicularis
12. *Lytocaryum weddellianum* (Dwarf coconut palm, Sago palm, Weddel palm)
13. *Normanbya normanbyi* (Black palm)
14. *Pinanga coronata* (Bunga, Ivory cane palm, Pinang palm)
15. *Ptychosperma macarthurii* (Hurricane palm, Macarthur feather palm, Macarthur palm)
16. *Reinhardtia gracilis* (Window palm, Window pane palm)
17. *Roystonea regia* (Cuban royal , Cuban royal palm, Florida royal palm, Royal palm)

Palms for subtropical and warm - temperate regions

1. *Arenga engleri* (Dwarf sugar palm, Sugar palm, Formosa palm)
2. *Brahea armata* (Blue fan palm, Blue hesper palm, Grey goddess, Mexican blue fan palm, Mexican blue palm, Short blue hesper)
3. *Butia capitata* (Butia palm, Jelly palm, Pindo palm, South American jelly palm, Wine palm)
4. *Chamaedorea elegans* (Good luck palm, Parlor palm, Parlour palm)
5. *Chamaedorea metallica* (Metallic palm, Miniature fishtail plam)
6. *Chamaerops humilis* (European fan palm, Fan palm, Mediterranean fan palm)
7. *Hedyscepe canterburyana* (Big mountain palm, Umbrella palm)
8. *Howea belmoreana* (Belmore sentry palm, Curly palm, Sentry palm)
9. *Howea forsteriana* (Forster sentry palm, Kentia palm, Paradise palm, Sentry palm, Thatch leaf palm)
10. *Linospadix monostachya* (Walking stick palm)
11. *Livistona australis* (Australian cabbage palm, Australian palm, Australian fan palm, Cabbage palm, Fan palm and Gippsland palm)
12. *Rhapis excelsa* (Bamboo palm, Fern rhapis, Ground rattan, Lady palm, Little lady palm, Slender lady palm, Miniature fan palm)
13. *Rhapis humilis* (Reed rhapis, Slender lady palm)
14. *Trachycarpus fortunei* (Chinese windmill palm, Chusan palm, Fan palm, Hemp palm, Windmill palm)

Palms as ground covers

1. *Arenga caudata* (Dwarf sugar palm)
2. *Chamaedorea cataractarum* (Cascade palm, Cat palm, Cataract palm)
3. *Chamaedorea elegans* (Good luck palm, Parlor palm, Parlour palm)
4. *Chamaedorea metallica* (Metallic palm, Miniature fishtail plam)
5. Chamaedorea stolonifera
6. *Reinhardtia gracilis* (Window palm, Window pane palm)
7. *Sabal minor* (Blue palmetto palm, Bush palmetto, Dwarf palm, Dwarf palmetto, Dwarf palmetto palm, Little blue stem, Scrub palmetto, Swamp palmetto)

Palms for hedging and screening purposes

1. *Acoelorraphe wrightii* (Everglades palm, Paurotis palm, Saw cabbage palm, Silver saw palm, Silver saw palmetto)
2. *Arenga caudata* (Dwarf sugar palm)
3. *Arenga engleri* (Dwarf sugar palm, Sugar palm, Formosa palm)
4. *Bactris gasipaes* (Chonta, Peach palm, Pejibaya, Pejibeye, Pejivalle, Pewa, Pupunha)
5. *Caryota mitis* (Burmese fishtail palm, Clustered fishtail palm, Fishtil palm and Tufted fishtail palm)
6. *Chamaedorea cataractarum* (Cascade palm, Cat palm, Cataract palm)
7. *Chamaedorea microspadix* (Bamboo palm, Hardy bamboo palm)
8. *Chamaedorea seifrizii* (Bamboo palm, Reed palm)
9. *Chamaerops humilis* (European fan palm, Fan palm, Mediterranean fan palm)
10. *Dypsis lutescens* (Areca palm, Butterfly palm, Cane palm, Golden feather palm, Golden yellow palm, Madagascar palm, Yellow palm, Yellow butterfly palm)
11. *Metroxylon sagu* (Sago palm)
12. *Oncopserma tigillarium* (Nibung palm)
13. *Phoenix reclinata* (African date palm, African wild date palm, Senegal date palm)
14. *Pinanga coronata* (Bunga, Ivory cane palm, Pinang palm)
15. *Ptychosperma macarthurii* (Hurricane palm, Macarthur feather palm, Macarthur palm)
16. *Rhapidophyllum hystrix* (Blue palmetto, Creeping palmetto, Dwarf saw palmetto, Hedgehog palm, Needle palm, Porcupine palm, Spine palm, Vegetable porcupine)
17. *Rhapis excelsa* (Bamboo palm, Fern rhapis, Ground rattan, Lady palm, Little lady palm, Slender lady palm, Miniature fan palm)
18. *Serenoa repens* (Saw palmetto, Scrub palmetto)

Palms for dry soils

1. *Arenga pinnata* (Aren, Areng palm, Black fibre palm, Gomuti palm, Kabong, Sugar palm)
2. *Bismarckia nobilis* (Bismarck palm)
3. *Borassus flabellifer* (Doub palm, Lontar plam, Palmyra palm, Pannamaram, Tala palm, Talauriksha palm, Tal-gas, Toddy palm, Wine palm)
4. *Brahea armata* (Blue fan palm, Blue hesper palm, Grey goddess, Mexican blue fan palm, Mexican blue palm, Short blue hesper)
5. *Butia capitata* (Butia palm, Jelly palm, Pindo palm, South American jelly palm, Wine palm)
6. *Chamaerops humilis* (European fan palm, Fan palm, Mediterranean fan palm)
7. *Coccothrinax argentata* (Florida silver palm, Silver palm, Silver thatch palm, Silvertop, Silver top palm)
8. *Coccothrinax crinita* (Mat palm, Old man palm, Old man thatch palm, Palma petate, Thatch palm)
9. *Copernicia macroglossa* (Cuban petticoat palm, Jata de Guanbacoa, Petticoat palm)
10. *Jubaeopsis caffra* (Pondoland palm)
11. *Latania loddigesii* (Blue latan, Blue latan palm)
12. *Latania lontaroides* (Red latan, Red latan palm)
13. *Latania verschaffeltii* (Yellow, Yellow latan palm)
14. *Livistona chinensis* (Chinese fan palm, Chinese fountain palm, Fan palm, Footstool palm, Fountain palm)
15. *Phoenix reclinata* (African date palm, African wild date palm, Senegal date palm)
16. *Phoenix rupicola* (Cliff date palm, East India wine palm, India date palm, Wild date palm)
17. *Pseudophoenix sargentii* (Buccaneer palm, Cherry palm, Florida cherry palm)
18. *Sabal minor* (Blue palmetto palm, Bush palmetto, Dwarf palm, Dwarf palmetto, Dwarf palmetto palm, Little blue stem, Scrub palmetto, Swamp palmetto)

19. *Sabal palmetto* (Blue palmetto, Cabbage palm, Cabbage palmetto, Cabbage tree, Common Palmetto, Palmetto, Palmetto palm)
20. *Serenoa repens* (Saw palmetto, Scrub palmetto)
21. *Thrinax parviflora* (Broom palm, Florida thatch palm, Iron thatch, Mountain thatch palm, Jamaica thatch palm, Palmetto thatch, Thatch, Thatch pole)
22. *Thrinax radiata* (Florida thatch palm, Jamaica thatch, Sea thatch, Silk-top tahtch)
23. *Trithrinax acanthacoma* (Buriti palm, Spiny fibre palm)

Palms for wet and swampy soils

1. *Acoelorraphe wrightii* (Everglades palm, Paurotis palm, Saw cabbage palm, Silver saw palm, Silver saw palmetto)
2. *Copernicia macroglossa* (Cuban petticoat palm, Jata de Guanbacoa, Petticoat palm)
3. *Cyrtostachys renda* (Sealing wax palm, Lipstick palm, Maharajah palm, Pinang-rajah)
4. *Elaeis guineensis* (African oil palm, Jacara palm)
5. *Euterpe edulis* (Assai palm, Jacara palm)
6. *Hydriastele wendlandiana* (Florence falls palm, Latrum palm)
7. *Livistona australis* (Australian cabbage palm, Australian palm, Australian fan palm, Cabbage palm, Fan palm and Gippsland palm)
8. *Metroxylon sagu* (Sago palm)
9. *Phoenix roebelenii* (Dwarf date palm, Miniature date palm, Pygmy date palm, Roebelin palm)
10. *Ptychosperma macarthurii* (Hurricane palm, Macarthur feather palm, Macarthur palm)
11. *Raphia farinifera* (Madagascar raffia palm, Latrum palm)
12. *Ravenea rivularis* (Majesty palm)
13. *Rhapidophyllum hystrix* (Blue palmetto, Creeping palmetto, Dwarf saw palmetto, Hedgehog palm, Needle palm, Porcupine palm, Spine palm, Vegetable porcupine)
14. *Roystonea regia* (Cuban royal , Cuban royal palm, Florida royal palm, Royal palm)

Palms for coastal regions

1. *Copernicia macroglossa* (Cuban petticoat palm, Jata de Guanbacoa, Petticoat palm)
2. *Phoenix dactylifera* (Date, Date palm)
3. *Pseudophoenix sargentii* (Buccaneer palm, Cherry palm, Florida cherry palm)
4. *Serenoa repens* (Saw palmetto, Scrub palmetto)
5. *Thrinax radiata* (Florida thatch palm, Jamaica thatch, Sea thatch, Silk-top thatch)

II. Choose the best answer

1. An example for fan shaped palm is

   Chamaerops humilis *Howea belmoreana* *Lytocaryum weddelianum*

2. An example for feather leaved palm is

   Trachycarpus fortune *Chamaerops humilis* *Archontophoenix alexandrae*

3. Fishtail palm is

   Howea belmoreana *Caryota mitis* *Trachycarpus fortune*

4. Entire-leaved palm is

   Chamaerops humilis *Archontophoenix alexandrae* *Johannesteijsmannia altifrons*

5. Trunkless palm is

   Sabal minor *Chamaerops humilis* *Archontophoenix alexandrae*

III. Match the following

1.	Clumping palm	:	*Hyphaena thebaica*
2.	Solitary palm	:	*Phoenix dactylifera*
3.	Branching palm	:	*Adonidia merrillii*
4.	Climbing palm	:	*Chamaedorea seifrizii*
5.	Bamboo palm	:	*Calamus australis*
6.	*Dypsis lutescens*	:	Dry soils
7.	*Coccothrinax crinita*	:	Wet and swampy area

8. *Thrinax radiata* : Screening
9. *Reinhardtia gracilis* Coastal region
10. *Cyrtostachys renda* Ground cover

Answers

II. Choose the best answer

1. *Chamaerops humilis*
2. *Archontophoenix alexandrae*
3. *Caryota mitis*
4. *Johannesteijsmannia altifrons*
5. *Sabal minor*

III. Match the following

1. Phoenix dactylifera
2. Adonidia merrillii
3. Hyphaena thebaica
4. Calamus australis
5. *Chamaedorea seifrizii*
6. Screening
7. Dry soils
8. Coastal region
9. Ground cover
10. Wet and swampy area

CYCADS

Superficially, cycads resemble palms and many plant enthusiasts are attracted equally to both. Botanically, however, they are quite distinct not only from palms, but from every other plant group and represent a plant order unrelated to any other. They are much older and more primitive than palms and appeared on the planet some 200 million years ago. Related more to conifers than any other plants, they produce cones, and male and female ones are on separate plants. Pollinatin is carried out either by wind or by insects and, if successful, results in the production of large seeds. They belong to the family cycadaceae.

S. No.	Botanical Name	Remarks
1.	*Bowenia serrulata*	Leaves are doubly pinnate, the small leaflets roughly diamond shaped with pointed tips, and glossy green with toothed margins, a key identification characters. The large under ground rootstock has many crowns and may be divided to propagate the plant. It grows to only about 1-2 m in height and the spreading leaves can be up to 6½ feet (2m) long and 3 feet (1m) wide. The seeds are 2 or 3cm long, oval in shape, and creamy white when ripe.
2.	*Bowenia spectabilis*	It too has an underground stem, in this case shaped like a carrot with a small number of crowns from it. The bipinnate leaves can be up to 6½ feet (2m) long by 3 feet (1m) wide, the diamond shaped leaflets are glossy, and the leaf edges are smooth as opposed to toothed. Fresh seeds should be stored for two to three months to enable them to ripen fully and soaked for 24 hours prior to planting.
3.	*Ceratozamia mexicana*	The pinnate leaves are up to 3 feet (1m) long and dark green, the leaflets are narrow and pointed at the tip. PLder plants may form a short trunk ot this may be subterranean. Plants grow to about 5 to 6½ feet (1.5 to 2m) tail and the leaves are held upright. The cones are cylinder-shaped; male and female are on separate plants, and the seeds are pinkish-white when ripe.

4.	*Cycas circinalis*(Fen palm)	The foliage is graceful and attractive, suitable for large containers and outdoor gardens.
5.	*Cycas panzhihuaensis*	It grows to about 6½ feet (2m) tall, and the trunk is some 12 inches (30cm) in diameter. The leaves are dark bluish-green and held flat. The oval seeds are an inch (2.5cm) long and red when ripe.
6.	*Cycas pectinata*	It has a lumpy trunk up to 40 feet (12m) in height, sometimes branched, covered with the scars of old leaf bases. Some of the taller trunks must be extremely old. The leaves are mid-green, 5 feet (1.5m) long by about 12 inches (30cm) wide, and the leaflets are narrow and closely spaced. The seeds are large, about 2 inches (5cm) long, oval in shape, and bright orange when ripe.
7.	*Cycas revolute*(Sago palm)	The most common and best known cycad, it is widely available in many nurseries. It can grow to 10 feet (3m) or more tall with a branching trunk approximately 12 inches (30cm) in diameter. The leaves are flatly held, 3 feet (1m) long and 8 inches (20cm) wide, and the leaflets are glossy dark green and closely spaced. The female cone is large and spherical, the male long and elongated. The seeds are red when ripe and about 1 inch (3cm) long, and oval in shape.
8.	*Dioon edule*	Solitary trunk grows ultimately to some 10 feet (3m) tall by about 12 inches (30cm) diameter, usually unbranched. The leaves are blue-green and the stiff, narrow, pointed leaflets are so closely spaced as to be sometimes overlapping, presenting a distinctive and pleasing appearance. The male cone is cylindrincal, the female oval in shape, and the seeds white when ripe.
9.	*Dioon spinulosum*	It has a stout trunk about 16 inches (40cm) in diameter and 30 feet (10m) or more tall. Specimens with 50 feet (15m) of trunk have been reported. The leaves comprise a rounded crown and are glossy green, about 3 or 6½ feet (1 or 2m) long, and arching. The leaflets, as the name suggest, are bordered with sharp spines, five or six along each side.

10. *Encephalartos bubalinus*

It is a small species from Tanzania, which grows to only about 6½ feet (2m) tall. The trunk is about 16 inches (40cm) in diameter and produces basal suckers which may be removed for propagation. The glossy, mid-green leaves are 3 feet (1m) long by 8 or 10 inches (20 or 30cm) wide and are arranged in a hemispherical crown. The male cones are tall and tapering, like a slim pine cone, the female are barrel shaped, green, and sit horizontally in the crown of leaves. The seeds are yellow or orange when ripe.

11. *Encephalartos lehmannii*

The blue form is perhaps the bluest of the all cycads, its smooth, spineless leaflets give it an attractive look and it is popular as a houseplant. There is also a green leaved form, less showy but nice-looking nonetheless. It has a trunk to about 6½ feet (2m) high by about 12 or 16 inches (30 or 40cm) in diameter.

12. *Encephalartos manikensis*

This attractive cycad has a short thick trunk to only about 6½ feet (2m) in height. The thorny, glossy, dark green leaves grow in a rounded crown. Male and female cones are on separate plants, and similar in shape, though the female is much shorter and thicker then the male. They are rather variable in size and occur in shades of green. The seeds, about 1 inch (3cm) long, are oval and red when ripe.

13. *Lepidozamia hopei*

It is a handsome plant with a 20 inch (50cm) diameter trunk that grows up to 70 feet (20m) in height, making it among **the tallest growing cycads** in the world. The dark green, glossy leaves can be 6½ or 10 feet (2 or 3m) long by up to 20 inches (50cm) wide, finely pinnate, without spines, and the leaflets closely spaced. The dissimilar cones are one per stem; the male cylindrical, the female oval in shape. The seeds are oblong, about 2 inches (5cm) long by ¾ inch (2cm) wide, and they are bright red.

14. *Macrozamia communis*

Usually the trunk is absent or below ground, but very old plants can grow a solitary trunk up to 6½ feet (2m) in height. The leaves are plentiful and finely pinnate, and form a rounded, elegant soft-looking crown. The male cones are cylindrincal; the female cones barrel-shaped, both blue-green, and there may be more than one cone per stem.

15. *Macrozamia moorei*

A stunning species with a single, massive trunk up to 3 feet (1m) in diameter and 20 feet (7m) in height, which is covered with countless thousands of diamond-shaped scars, each one representing a fallen leaf. The male cones are cylindrical and numerous, the female cones wider and thicker, just a few per stem. The seeds are oval, up to 2 inches (5cm) long, and bright red when ripe.

16. *Zamia furfuracea* (Cardboard palm) *Z. furfuracea* 'varigata'

Named for its tough and leathery leaves, the Cardboard zamia is widely available as a pot or houseplant, and as such performs well. It has a branching underground stem which produces several crowns of leaves. These have roughly diamond-shaped leaflets, thick and leathery. In sunny situations these are really crowded and ovrlap each other. The male cone is cylindrical; the female is barrel-shaped and produces small, bright red seeds.

17. *Zamia pseudoparasitica*

It has a small trunk with 3 to 10 feet (1 to 3m) long leaves, with glossy, mid-green, pointed leaflets, hang down from the tree. The cones are similar, with the male being narrower than the female and these are upright within the crown. The seeds are oval, ¾ inch (2cm) long, and yellow when ripe.

Chapter **9**

Bulbous Plants

I. General information

1. Plants propagated through modified underground stems are grouped into **tubers, bulbs, rhizomes**
2. An example of a bulbous plant propagated through **tuber** is **dahlia**
3. Bulbous plants which are propagated through **corms** are *Gladiolus, Freesia, Tetonia and Tigrida*
4. Bulbous plants propagated through **bulbs** are *Amaryllis, Tuberose, Zephyranthes* and *Crinum*
5. Bulbous plants propagated through **rhizome** are ***Alstroemeria,*** **Bamboo,** ***Canna, Convallaris, Hydechium, Iris*** **and** ***Zantedeschia***
6. Bulbous plant propagated through **tubercle** is ***Bowiea volubilis***
7. Bulbous plants propagated through **tuberous root**s are ***Dahlia, Eranthis, Ipomoea, Mirabilis, Ranunculus*** **and** ***Tapioca***
8. Bulbous plants propagated through **tuberous stem** are ***Begonia*** **and** ***Cyclamen***
9. Bulbous plant propagated through **offsets** are **daffodil, tulip and iris**
10. Bulbous plant propagated through **bulblet** is ***Lilium***
11. Bulbous plant propagated through **bulbil** is ***Lilium***
12. Bulbous plant propagated through **stem cutting** is ***Lilium***
13. Bulbous plants propagated through **scales** are ***Lilium*** and ***Amaryllis***
14. Bulbous plants propagated through **basal plate cut pieces** are **Amaryllis, Day lily and Hyacinth**

15. Bulbous plant propagated through **cut pieces of bulb** are **Amaryllis, Hyacinth and Tulip**
16. Bulbous plant propagated through **leaf cuttings** are **Football lily, Hyacinth and Muscari**
17. Bulbous plants which are grown to create mass effect are ***Dahlia, Canna*** **and** ***Zephyranthes***
18. Bulbous plants which are cultivated as cut flowers are ***Gladiolus, Tuberose*** **and** ***Amaryllis***
19. Bulbous plants which are grown as potted plants are ***Amaryllis and Dahlia***
20. Bulbous plant which is cultivated for extraction of essential oil is **Tuberose**
21. Modified underground stems of bulbous plants are left in the ground after harvesting of flowers belongs to **hardy** types
22. Hardy type bulbous plants are ***Amaryllis,*** **Tuberose,** ***Canna, Zephyranthes, Crinum,*** **etc**
23. Modified underground stems of bulbous plants which are lifted from the ground soon after harvesting of flowers are of **tender** types
24. Tender type bulbous plants are ***Gladiolus, Narcissus, Freesia, Daffodils*** **and** ***Dahlia***
25. Bulbous plants which are suitable to grow under warm climate are **Tuberose,** ***Canna, Zephyranthes, Crinum*** **and** ***Dahlia***
26. Bulbous plants which are suitable to grow under cool climate are **Tuberose** ***Gladiolus, Narcissus, Freesia*** **and** ***Daffodils***
27. Bulbous plants which are propagated through terminal cuttings are **Dahlia and Rex begonia**
28. Bulbous plant in which the scales closly packed the bulb is ***Hyacinthus***
29. Bulbous plants in which the scales loosely packed the bulb is **Lily**
30. Bulbous plants suitable for **shade** are

Allium ursinum	*Cyclamen purpurascens*
Anemone blanda	*Disporum flavescens Erythronium*
Anemone nemorosa	*Fritillaria pallidiflora*

Arum italicum
Convallaria majalis
Corydalis flexuosa
Iris douglasiana
Lilium martagon
Sanguinaria canadensis
Trillium spp.
Galanthus spp.
Hyacinthoides non-scripta
Leucojum vernum
Ranunculus ficaria
Smilacina racemosa
Uvularia grandiflora

31. Bulbous plants suitable for **rock garden** are

Allium spp.
Anemone spp.
Anthericum spp.
Bulbocodium spp.
Galanthus spp.
Ipheion spp.
Muscari spp.
Ornithogalum spp.
Chionodoxa spp.
Cyclamen spp.
Eranthis spp.
Erythronium spp.
Romulea spp.
Rhodohypoxis spp.
Scilla spp.
Oxalis spp.

32. **Summer flowering** bulbous plants are

Achimenes spp.
Agapanthus spp.
Allium spp.
Alstroemeria spp.
Amaryllis spp.
Anemone spp.
Begonia spp.
Calochortus spp.
Haemanthus spp.
Lilium spp.
Oxalis spp.
Tritonia spp.
Watsonia spp.
Hymenocallis spp.
Canna spp.
Crocosmia spp.
Dahlia spp.
Dierama spp.
Eucomis spp.
Galtonia spp.
Gladiolus spp.
Gloriosa spp.
Pancratium spp.
Pardisea spp.
Polianthes spp.
Sprekelia spp.
Zantedeschia spp.

33. Bulbous plants which are of **spring flowering** types are

Allium spp.	*Hyacinthus* spp.
Arum spp.	*Hippeastrum* spp.
Asphodelus spp.	*Iris* spp.
Camassia spp.	*Ixia* spp.
Convallaria spp.	Leucojum spp.
Crocus spp.	*Muscari* spp.
Cyclamen spp.	*Narcissus* spp.
Eranthis spp.	*Ornithogalum* spp.
Freesia spp.	*Ranunculus* spp.
Fritillaria spp.	*Scilla* spp., Trillium spp.
Galanthus spp.	*Zephyranthes* spp.

34. Optimum time for planting bulbous plants is **October to December**

35. Planting depth of following bulbous plants are

Anemone, Anemonella, Claytonia, Corydalis, Dentaria, Eranthis	2.5 cm
Brodiaea, Ipheion, Cyclamen	5-7.5 cm
Allium, Arisaema, Bellevallia, Brimeura, Bulbocodium, Chinodoxa, Crocus, Cypripedium, Dicentra, Endymion, Erythronium, Fritillaria, Galanthus, Hycinthoides, Iris danfordiae, Iris reticulata, Muscari, Oxalis, Ornithogalum, Puschkinia, Scilla, Trillium	7-10 cm
Colchicum, Iris bucharica, Eremus, Leucojum, Liatris, Narcissus, Tulipa	10-12 cm
Allium, Camassia, Fritillaria, Hyacinthus, Iris hybridum, Iris iphioides, Lilium, Narcissus, Tulipa	15-20 cm

36. Bulbous plants tolerant to **moist conditions** are

Allium	*Crocus*
Anemone	*Endymion*
Brimeura	*Erythronium*
Camassia	*Galanthus*
Chinodoxa	*Leucojum*
Lilium	*Lycoris*

Muscari	*Narcissus*
Scilla	*Trillium*

37. Bulbous plants which possess tunicate and concentric bulbs are *Amaryllis,* daffodils, Hyacinth, Narcissus, Nerine and Tulip
38. Bulbous plants which possess non tunicate bulbs and non concentric bulbs is lilium
39. Country which is leading in bulbous plant cultivation is South Africa
40. In India, Kalimpong and Sikkim are highly concentrated in cultivating bulbous plants in large area
41. Ideal planting time for tuberose and zephyranthes is February and March
42. Ideal planting time for canna is July
43. Ideal planting time for Gladiolus, Narcissus and Daffodils is mid September to mid November
44. Ideal planting time for spring flowering bulbous plants is September to November
45. Ideal planting time for summer flowering bulbous plants is February to April
46. Ideal planting time for autumn flowering bulbous plants is July to August
47. Bulbous plants possess the growth cycle of flowering-vegetative-dormancy are football lily, zephyr lily, red spider lily and colchicum
48. Bulbous plants which require hot-cool- warm thermo condition are canna, crinum, dahlia, tuberose and zephyr lily
49. Bulbous plants which require warm-cool-warm thermoperiodic condition are daffodil, freesia, gladiolus and narcissus
50. Tulip bulbs when they attain 'G' stage needs cold treatment
51. Tulip breaking virus is a major disease in tulip cultivation
52. Tulip breaking virus is transmitted by aphids
53. Country which is the leading exporter of bulbs is the Netherlands
54. Bulbous plants which are native of India are crinum, gloriosa, begonia

55. Evergreen bulbous plants are clivia, kniphofia, dietes
56. Deciduous bulbous plants are crocus, tulip and watsonia
57. A state which is highly suitable for growing bulbous plants is Himachal Pradesh
58. Ideal temperature for storing bulbous plants to putforth normal growth is 18-20^{0}C
59. While forcing the bulbs for early flowering, bulbs can be stored at 30-35^{0}C temperature
60. Ideal temperature for storing crocus bulbs is 10-12^{0}C
61. Common name for the bulbous plants is

Amaryllis	Belladona lily, king or star lily
Belamcanda chinensis	Leopard Flower, Blackberry lily, Belamcanda
Caladium spp.	Heart of Jesus, Angel-wings
Canna indica	Indian shot
Crinum bulbispermum	St. John lily, English Cape lily, Swamp lily
Gerbera jamesonii	Transvaal Daisy, African daisy
Globba marantina	Dancing girl ginger/yellow dancing girl
Haemanthus multiflorus	Blood lily, Red cape lily, Football lily
Hedychium spp.	Ginger lily
Alpinia speciosa	Shell ginger
Hemerocallis fulva	Day lily
Hymenocallis littoralis	Spider lily
Iris sp.	Flag flower
Narcissus sp.	Nargis
Polianthes tuberosa	Rajani Gandha, Tuberose
Zantedeschia spp.	Calla lily
Zephyranthes sp.	Zephyr lily

II. Choose the best answer

1. *Haemanthes multiflorus* is commonly called as..........................

   Zephyr lily   Blood lily   Flag flower

2. Example for evergreen bulbous plant is..........................

   Kniphofia   Crocus   Tulip

3. Example for summer flowering bulbous plant is..........................

   Agapanthus spp.   *Cyclamen* spp.   *Iris* spp.

4. Bulbous plants propagated through tubercles is..........................

   Bowiea volubilis   *Alstroemeria*   Bamboo

5. Bulbous plant which is propagated by bulb is..........................

   Amaryllis   *Dahlia*   *Eranthis*

6. Bulbous plant which is cultivated for cut flower is..........................

   Gladiolus   *Dahlia*   Canna

7. Bulbous plant propagated through terminal cutting is...................

   Gladiolus   *Dahlia*   Canna

8. Bulbous plant which is native of India is..........................

   Gloriosa   *Amaryllis*   *Dahlia*

9. Bulbous plant which possess non tunicate bulbs and non concentric bulb is..........................

   Crocus   Tulip   Lilium

10. Spring flowering bulbous plant is..........................

    Freesia spp.   *Lilium* spp.   *Gladiolus* spp.

III. Fill in the blanks

1. Ideal planting time for canna is..........................
2. A state in India which is highly suitable for growing bulbous plants is..........................
3. Country which is the leading exporter of bulbs is..........................
4. Country which is leading in bulbous plant cultivation is..........................

5. Ideal planting time for summer flowering bulbous plants is...........................
6. Major disease of tulip cultivation is...........................
7. Ideal temperature for storing crocus bulbs is...........................
8. A stage in which tulip bulbs needs cold treatment is...........................
9. Bulbous plant in which the scales are closely packed the bulb is...........................
10. Bulbous plant in which the scales are closed packed the bulbs is ....

Answers

II. Choose the best answer

1. Blood lily
2. Kniphofia
3. *Agapanthus* spp.
4. *Bowiea volubilis*
5. *Amaryllis*
6. *Gladiolus*
7. Dahlia
8. Gloriosa
9. Lilium
10. *Freesia* spp.

III. Fill in the blanks

1. July
2. Himachal Pradesh
3. Netherlands
4. South Africa
5. February to April
6. Tulip breaking virus
7. 10-12^{0}C
8. 'G'stage
9. *Hyacinthus*
10. Lily

ANNEXURE VI

BULBOUS PLANTS

S. No.	Botanical Name	Common name	Family	Flowering period	Remarks
1.	*Achimene longiflora*	Magic flower	Gesneriaceae	June-November	Pot plant
2.	*Acidanthera bicolor* var.Murielae	Scented gladiolus	Irudaceae	February - April	Fragrance
3.	*Agapanthus umbellatus*	Blue African lily	Liliaceae	July - September	Pot plant
4.	*Allium species*	Allium	Alliaceae	May - October	Pot plant
5.	*Alpinia speciosa*	Shell ginger	Zingiberaceae	-	-
6.	*Alstroemeria sp.*	Incas lily	Alsteroemeriaceae	February - May	Cut flower and pot plant
7.	*Amaryllisbelladonna*	Belladonna lily / King or Star lily	Amaryllidaceae	April - June	Pot Plant
8.	*Anemone coronaria*	Wind flower	Ranunculaceae	February - April	Bedding plant
9.	*Begonia hydrida*	Tuberous begonia	Begoniaceae	April - August	Pot plant
10.	*Begonia rex*	Rex begonia	Begoniaceae	-	Foliage beauty
11.	*Begonia semperflorens*	Fibrous rooted begonia	Begoniaceae	Summer	Pot plant
12.	*Belamcanda chinensis*	Leopard flower	Iridaceae	August - October	Bedding plant
13.	*Canna indica*	Indian shot	Cannaceae	Through-out year	Bedding plant
14.	*Clivia indica*	Kaffir lily	Amaryllidaceae	February - April	Pot plant
15.	*Convallaria majalis*	Lily of valley	Liliaceae	March - June	Pot plant

16.	*Crinum bulbispermum*	Cape lily	Amaryllidaceae	July - September	Bedding plant
17.	*Crocosmia aurea*	Montbretia	Iridaceae	July - September	Bedding plant and cut flower
18.	*Crocus sativus*	Saffron	Iridaceae	September - December	Bedding and pot plant in hills only
19.	*Cyclamen persicum*	Cyclamen	Primulaceae	December - May	Pot plant
20.	*Dahlia variabilis*	Dahlia	Asteraceae	December-August	Bedding and pot plant
21.	*Eremurus himalaicus*	Foxtail lily	Liliaceae	April - July	Bedding and cut flower
22.	*Eucharis amazonica*	Eucharis	Amaryllidaceae	February - April	Pot plant
23.	*Fressia refrecta*	Fressia	Iridaceae	December – April	Pot plant
24.	*Fritillaria imperialis*	Crown imperial	Liliaceae	February - May	Pot plant
25.	*Gerbera jamesonii*	Transvaal Daisy/ Africa Daisy	Compositae	May - September	Cut flower and pot plant
26.	*Gloriosa superba*	Glory lily	Liliaceae	July - September	Bedding plant
27.	*Gloxinia speciosa*	Gloxinia	Gesneriaceae	May - July	Pot plant
28.	*Haemanthes multiflorus*	Blood flower/ Red cape lily	Amaryllidaceae	May - September	Pot plant
29.	*Hemerocallis fulva*	Day lily	Liliaceae	May - August	Bedding plant
30.	*Hippeastrum species*	Hippeastrum	Amaryllidaceae	March - May	Bedding, border and pot plant
31.	*Hyacinthus orientalis*	Hyacinth	Liliaceae	February - May	Bedding plant
32.	*Hymenocallis litterolis*	Spider lily	Amaryllidaceae	April - August	Bedding plant
33.	*Iris nepalensis*	Flag flower	Iridaceae	February - April	Bedding plant and cut flower

34.	*Kniphofia aloides*	Torch lily	Liliaceae	March - August	Bedding plant and cut flower
35.	*Lilium bulbiferum*	Orange lily	Liliaceae	March – July	Bedding plant and cut flower
36.	*Lilium candidum*	Madonna lily	Liliaceae	April – July	Bedding plant and cut flower
37.	*Lilium longiflorum*	Easter lily	Liliaceae	March – July	Bedding plant and cut flower
38.	*Lilium tigrinum*	Tiger lily	Liliaceae	April – July	Bedding plant and cut flower
39.	*Moraea iridiodes*	Moraea	Iridaceae	March - May	Bedding plant
40.	*Muscari botryoides*	Grape hyacinth	Liliaceae	February - May	Pot plant
41.	*Narcissus pseudonarcissus*	Daffodil	Amaryllidaceae	December - April	Cut flower and pot plant
42.	*Narcisssus tazetta*	Nargis	Amaryllidaceae	December - April	Cut flower and pot plant
43.	*Nelumbo lutea*	American lotus	Nymphaeceae	February - July	Ponds
44.	*Nelumbo nucifera*	Indian lotus	Nymphaceae	February - July	Ponds
45.	*Nerine sarniensis*	Nerine	Amaryllidaceae	August - October	Pot plant
46.	*Nymphaea lotus*	Water lily	Nymphaeceae	March - August	Ponds
47.	*Ornithogalum thyrsoides*	Chincheerinchee	Liliaceae	April - June	Bedding plant and cut flower
48.	*Oxalis species*	Oxalis	Oxalidaceae	February - June	Bedding plant and hanging basket
49.	*Phaedranassa schizantha*	Queen lily of Ecuador	Amaryllidaceae	March - May	Bedding plant
50.	*Polianthes tuberosa*	Tuberose	Amaryllidaceae	Throughout year	Bedding plant, cut and loose flower
51.	*Ranunculus asiaticus*	Ranunculus	Ranunculaceae	June - October	Bedding plant

52.	*Scilla hispanica*	Wild hyacinth	Liliaceae	April - June	Pot plant
53.	*Sparaxix tricolor*	Wand flower	Iridaceae	February - July	Bedding plant
54.	*Sprekelia formosissima*	Jacobaean lily	Amaryllidaceae	July - September	Pot plant
55.	*Tigridia pavonia*	Tiger Iris	Iridaceae	July - August	Pot plant
56.	*Tulbaghia violacea*	Tulbaghia	Liliaceae	March - May	Bedding plant
57.	*Tulipa species*	Tulip	Liliaceae	February - June	Cut flower, pot plant and bedding plant
58.	*Vallota purpurea*	Scarbo-rough lily	Amaryllidaceae	March - July	Pot plant
59.	*Watsonia species*	Watsonia	Iridaceae	March - May	Bedding plant
60.	*Zantedeschia aethiopica*	Calla/Arum lily	Araceaec	February - June	Bedding plant and cut flower
61.	*Zephyranthes grandiflorus*	Zephyr lily	Amaryllidaceae	April - July	Pot plant, bedding plant and rock gardens

Chapter **10**

Cacti & Succulents

I. General information

1. All cacti are succulents
2. All succulents are not cacti
3. All the genus belonging to the family Cactaceae are native of America
4. A genus in Cactaceae family which is not native of America is *Rhipsalis*
5. Sedum and Sempervivum are native to temperate Europe
6. *Aporocactus, Rhipsalis* and *Schlumbergera* are native of Mexico
7. *Dischidia* and *Hoya* are native of South-East Asia
8. *Cissus quadrangularis* is native of Africa
9. Climbing *Epiphyllum, Hylocereus* and *Selenicereus* are native of America
10. Cacti and succulents suitable for rock garden are

Adenium	Senecios	Haworthias
Aloes	Euphorbias	Kalanchoes
Agaves	Echinocactus	Mammillarias
Beaucarneas	Echinopsis	Pachypodiums
Ceropegias	Ferocactus	Yuccas
Sedums	Furcraeas	Crassulas
Gasterias		

11. Succulents used as house plants are

- *Beaucarnia recurvata*
- *Dracaena draco*
- *Yucca glauca*
- *Haworthias*
- *Senecios*

12. Succulents which are grown in pots are
   - ❐ *Beaucarnia recurvata*
   - ❐ *Adenium*
   - ❐ *Sedum*
13. Succulents suitable for planting in hanging baskets are
   - ❐ *Sedum*
   - ❐ *Aporocactus*
   - ❐ *Ceropegia*
   - ❐ *Epiphyllum*
   - ❐ *Zygocactus*
   - ❐ *Senecio*
14. All members of cactaceae family produce areoles
15. Areoles are round or oval in structure composed of two perpendicular buds, from the upper bud, flowers, fruits or new branches develop, while from lower ones spines will arise
16. Genus of cactaceae family in which spines are replaced by bristles or hairs are
   - ❐ *Ariocarpus*
   - ❐ *Echinocactus*
   - ❐ *Lophophora*
   - ❐ *Zygocactus*
   - ❐ *Epiphyllum*
17. Cacti genus in which roots are swollen and shaped like carrots are
   - ❐ *Coryphantha*
   - ❐ *Lophophora*
   - ❐ *Ariocarpus*
   - ❐ Many species of *Mammillaria*
   - ❐ *Echinocactus*
18. Cacti genus producing tuberous root which resembles dahlia are *Pterocactus, Wilcoxia, Peniocereus*
19. Cacti genus which produces woody stems are *Pereskia* and *Peireskoiosis*
20. The genus of cacti which possess cephalium is *Melocactus*
21. Cephaliumis defined as cacti which are found with dense bristles and hairs
22. Malformation seen in cacti is due to physical factors caused due to the formation of crests and monstrosities
23. Many genus of cactaceae family when grafted on the rootstock of *Opuntia moliviensis* become 'Monstrosities'
24. Succulents with waxy layer as a protective device is seen in *Dudleya farinosa*

25. Succulents with white felt of woolly hairs are seen in *Senecio* and *Crassula*

26. Examples for stem succulents are *Euphorbia* and *Testudinaria*

27. Examples for root succulents are *Adenium obesum* and *Beaucarnia recurvata*

28. Succulents seen with **caudex formation** are *Adenium obesum* and *Beaucarnia recurvata*

29. Columnar shaped succulent is *Pachypodim lamerei*

30. Spherical shaped succulent is *Euphorbia obesa*

31. Succulents which develop windows are *Lithops, Fenestraria, Opthalmopyllum* and *Haworthia*

32. The term 'Windows' in succulents is defined as a tissue lacking chlorophyll to make access of light to the assimilating tissue which are situated at the tip of the leaves

33. Succulent which looks like pebbles is lithops

34. Succulents which are suitable for planting in outdoor conditions are *Agave, Adenium, Beaucarnia, Echinopsis, Gymnocalycium, Cereus, Opuntia*

35. Duration taken by *Mesembryanthemum* seeds to germinate is 3-60 days

36. Seeds of Asclepiadaceae (*Stapelia, Huernia, Tavaresia, etc*) can germinate within twenty four hours

37. Seeds of *Haworthia, Gasteria, Aloe* can germinate in 2-4 weeks

38. Seeds of *Ceropegia fusca, Beaucarnia gracilis* and *Agave lechuguila* can germinate in 1-2 weeks

39. Succulents propagated through leaf cuttings are *Kalanchoe, Sedum* and *Gasteria*

40. Cacti and succulents propagated through leaf bud are *Sedum, Gasteria, Haworthia, Kalanchoe, Peireskiopsis, Euphorbia decaryi*

41. Cacti which are propagated by offsets are *Rebutia, Mamillaria, Notocactus, Trichocereus, Cleistocactus, Echinocereus*

42. Cacti which are propagated by cleft or wedge grafting are *Aporocactus, Wilcoxia, Zygocactus*

43. The best season for multiplication of cacti and succulents by cuttings is February and March
44. While grafting the cactus, scion of *Rebutia senilis* shows compatibility on *Myrtillocactus geometrizans* rootstock
45. While grafting the cactus, scion of *Peireskiopsis* shows compatibility on *Gymnoculycium* and *Turbinicarpus* rootstock
46. While grafting the cactus, scion of *Selenicereus macdonaldiae* shows compatibility on *Wilcoxia viperiana* and *Wilcoxia schmollii*
47. While grafting the cactus, scion of *Phyllocactus* and *Zygocactus* shows compatibility on *Harrisia tortuosa* and *Opuntia ficus-indica*
48. Periclinal chimera is reported in the grafted plant of *Gymnocalycium mihanovichi friedrichii 'Rubra'* on the rootstock of *Hylocereus* or *Myrtillocactus* in Japan
49. Success of succulents seen in *Pachypodium lamerei* and *Pachypodium geayi* when grafted on *Pachypodium lamerei* rootstock
50. Success of succulents seen in *Didiera madagascariensis* when grafted on *Peireskiopsis vellutina* rootstock
51. *Adenium* is vegetatively propagated by side grafting
52. Cacti with long tap roots (Carrot-like roots) are planted in deep pots
53. Succulents which produce tap roots are *Lithops, Pachypodium and Monadanium*
54. *Euphorbias* are preferrable to plant in very deep, narrow pots
55. Shallow root producing succulents are Haworthias, Stapelias
56. Shallow rooted succulents are planted in shallow pots
57. Cacti suitable for planting in shallow pots are *Mammillaria, Echinocereus* and *Hamatocactus*
58. Media used for planting *Rhipsalis* is leafmould and loam
59. Media used for planting *Glottiphyllum* is leafmould, loam and sand
60. Diseases reported in succulents are dry rot, black rot, damping off, *Phytophthora, Fusarium* and *Rhizoctonia*
61. Pests reported in succulents are mealybugs, scale insects, spiders and nematodes

62. Common name for **cacti** is as follows

Aporocactus flagelliformis	:	Rat-tail cactus
Ariocarpus fissuratus	:	Living rock
Astrophytum asterias	:	Sea-urchin cactus
Astrophytum myriostigma	:	Bishop's cap, Bishop's hood
Astrophytum ornatum	:	Monk's hood
Bowea volubilis	:	Sea onion
Cephalocereus senilis	:	Old man cactus
Cereus spp.	:	Night blooming cacti
Cereus uruguayanus	:	Hedge blooming cacti
Crassula hemisphaerica	:	Arab's turban
Echinopsis spp.	:	Barrel or hedge hog cacti
Echinocactus grusonii	:	Golden barrel, Golden ball, Mother in laws chair
Lobivia famatimensis	:	Orange cob cactus
Lophophora williamsii	:	Peyote or whisky cactus, Mescal button
Mammillaria sp.	:	Fish hook or pincushion cacti or Bird's nest cacti
Mammillaria bocasana	:	Powderpuff
Mammillaria candida	:	Snow ball cactus
Mammillaria celsiana	:	Showy pincushion
Mammillaria hahniana	:	Old lady
Mammillaria compressa	:	Mother of hundred
Mammillaria rhodantha	:	Rainbow pincushion
Melocactus bahiensis	:	Turk's head
Notocactus leninghausii	:	Golden ball
Notocactus scopa	:	Silver ball
Opuntia clavarioides	:	Sea coral
Opuntia falcata	:	Tree opuntia
Opuntia ficus-indica	:	Indian fig

Opuntia microdasys : Bunny ears

Oreocereus celsianus : Old man of the Andes

Pereskia aculeata : Barbados gooseberry or lemon vine

Pereskia bleo : Wax rose

Pereskia grandiflora : Rose cactus

Gymnocalycium sp. : Chin cactus

Gymnocalycium mihanovichii var. Friederickii : Mr. Red cap

Lobivia sp. : Cob cacti

Rebutia minuscula : Red crown

Rhipsalis burchellii : Mistletoe

Thelocactus bicolor : Glory of Texas

Selenicereus sp. : Moon cactus

63. Common name for **succulents** is as follows

Adenium obesum 'Obesum' : Desert rose, Mock azalea

Agave americana : Century plant

Agave angustifolia : Caribbean agave

Agave filifera : Thread plant

Agave parviflora : Little princess agave

Euphorbia spp. : Mother in law's tongue

Agave potatorum : Drunkard agave

Agave striata : Dark striped agave

Agave striata 'Nana' : Liliput agave

Agave stricta Hedge hog agave

Agave victoriae-reginae : Queen agave

Aloe aristata : Lace aloe

Aloe humilis : Spider aloe

Aloe humilis 'Echinata' : Hedge hog aloe

Beaucarnea gracillis : Bottle palm

Beaucarnea recurvata	:	Elephant foot tree
Bowiea volubilis	:	Climbing onion or sea onion
Ceropegia woodii	:	String of hearts
Crassula arborescens	:	Silver dollar, Silver jade
Crassula falcata	:	Scarlet paint brush, airplane plant
Dasylirion longissimum	:	Mexican grass tree
Euphorbia grandicornis	:	Cow-horn euphorbia
Euphorbia fusiformis	:	Ban muli, Chepudu gadda
Euphorbia milii	:	Crown of thorns
Euphorbia obesa	:	Baseball plant
Euphorbia tirucalli	:	Milk bush
Euphorbia horrida	:	African milk barrel
Furcraea gigantia 'Medio-picta'	:	Giant false agave
Gasteria verrucosa	:	Warty aloe
Huernia zebrina	:	Owl eyes, zebra flower
Kalanchoe beharensis	:	Elephant ears
Kalanchoe tomentosa	:	Panda plant
Lithops spp.	:	Old man's tooth
Lithops bella	:	Pretty stone face
Lithops karasmontana	:	Mountain pebble
Pachypodium lealii	:	Madagascar palm, Club-foot
Phytolacca dioica	:	Elephant tree
Testudinaria macrostachya	:	Tortoise plant
Yucca sp.	:	Spanish day
Yucca gloriosa	:	Spanish dagger
Sedum stahlii	:	Beads

64. Cacti genus used as a rootstock for grafting are *Cereus, Trichocereus, Harrisia, Opuntia* and *Cleistocactus*

65. The ideal season for sowing cacti seeds is March to September

66. Tribes seen in cactaceae family are Pereskieae, Opuntieae and Cereae
67. Cacti belongs to Pereskieae family is *Pereskia grandiflora* and *P. aculeata*
68. Cacti belongs to Opuntieae family is *Opuntia* sp.
69. Subtribes of the tribe cereae in cactaceae are Cereanae, Hylocereanae, Echinocereanae, Echinocactaenae, Cactanae, Coryphanthanae, Epiphyelanae, Rhipsalidanae
70. Cacti belongs to cereanae subtribe are cereus, Cephalocereus, Cleistocactus, Trichocereus
71. Cacti belong to Hylocereanae subtribe are Selenicereus, Hylocereus, Weberocereus, Cryptocereus
72. Cacti belong to Echinocereanae subtribe are Echinocereus, Echinopsis, Rebutia, Chamaecereus

II. Fill in the blanks

1. A genus in Cactaceae family which is not native of America is ....
2. *Cissus quadrangularis* is native of ....
3. The genus of cacti which possess cephalium is..............
4. Columnar shaped succulents is ....
5. Spherical shaped succulents is ....
6. Succulent which looks like pebbles is ....
7. *Adenium* is vegetatively propagated by ....
8. The ideal season for sowing cacti seeds is ....
9. Cacti belong to Opuntieae family is ....
10. The best season for multiplication of cacti and succulents by cuttings is ....

III. Choose the best answer

1. Cactus which planted in shallow pots is ....

   Aporocactus spp. *Mammillaria* spp. *Cereus* spp.

2. Succulent which produce tap roots is ....

   Lithops *Agave* spp. *Crassula* spp.

3. *Cleistocactus* is commercially propagated by ....

   Grafting Offsets Seeds

4. *Kalanchoe* is commercially propagated by ....

   Grafting Offsets Leaf cuttings

5. Succulent which is seen with caudex formation is ....

   Bowiea volubilis *Beaucarnia recurvata* *Euphorbia tirucalli*

6. Example for stem succulent is ....

   Euphorbia *Bowiea volubilis* *Beaucarnia recurvata*

7. Succulent seen with waxy layer as a protective device is ....

   Senecio *Dudleya farinosa* *Crassula*

8. Monstrosities is observed in cacti genus when grafted on ....rootstock

   Opuntia moliviensis *Melocactus bahiensis* *Rebutia minuscula*

9. Leafy cactus is ....

   Aporocactus spp. *Mammillaria* spp. *Pereskia*

10. Cactus which possess carrot shaped roots is ....

    Ariocarpus *Aporocactus* spp. *Zygocactus*

IV. Match the following

1. Cactus : Snow ball cactus
2. *Ariocarpus fissuratus* : Sea coral
3. *Bowea volubilis* : Moon cactus
4. *Mammillaria candida* : Areoles
5. *Opuntia clavarioides* : Living rock
6. *Selenicereus* sp. : Sea onion
7. *Sedum stahlii* : Century plant
8. *Kalanchoe beharensis* : Beads
9. *Euphorbia tirucalli* : Elephant ears
10. *Agave americana* : Milk bush

Answers

II. Fill in the blanks

1. *Rhipsalis*
2. Africa
3. *Melocactus*
4. *Pachypodim lamerei*
5. *Euphorbia obesa*
6. Lithops
7. Side grafting
8. March to september
9. *Opuntia* sp.
10. February and march

III. Choose the best answer

1. *Mammillaria*
2. *Lithops*
3. Offsets
4. Leaf cuttings
5. *Beaucarnia recurvata*
6. *Euphorbia*
7. *Dudleya farinosa*
8. *Opuntia moliviensis*
9. *Pereskia*
10. *Ariocarpus*

IV. Match the following

1. Areoles
2. Living rock
3. Sea onion
4. Snow ball cactus
5. Sea coral
6. Moon cactus
7. Beads
8. Elephant ears
9. Milk bush
10. Century plant

ANNEXURE VII

CACTI AND SUCCULENTS

S.No.	Botanical Name	Common name
1.	*Acanthocalycium violaceum*	-
2.	*Aporocatcus flagelliforus*	Rat-tail cactus
3.	*Aricocarpus fissuratus*	Living rock
4.	*Aricocarpus kotschoubeyanus*	-
5.	*Aricocarpus retusus*	-
6.	*Aricocarpus trigonus*	-
7.	*Astrophytum asterias*	Sea-urchin cactus
8.	*Astrophytum capricorne*	-
9.	*Astrophytum myriostigma*	Bishop's cap
10.	*Astrophytum ornatum*	Monk's hood
11.	*Aztekium ritteri*	-
12.	*Blossfeldia liliputana*	-
13.	*Borzicatcus madisoniorum*	-
14.	*Borzicatcus samaipatanus*	-
15.	*Buiningia brevicylindrica*	-
16.	*Carnegiea gigantean*	Saguaro
17.	*Caphalocereus albispinus*	-
18.	*Cephalocereus palmerii*	-
19.	*Cephalocerus senilis*	Old man cactus
20.	*Cereus uruguyanas*	Hedge cactus
21.	*Chamaecereus silverstrii*	-
22.	*Cleistocatcus strausii*	-
23.	*Copiapoa humilis*	-
24.	*Copiapoa montata*	-
25.	*Coryphantha bumamma*	-
26.	*Coryphantha cornifera*	-
27.	*Coryphantha elephantidens*	-
28.	*Coryphantha radians*	-
29.	*Discocatcus araneispinus*	-
30.	*Discocatcus hartmannii*	-
31.	*Discocatcus horstii*	-

32.	*Dolichothele longimamma*	-
33.	*Dolichothele sphaerica*	-
34.	*Dolichothele uberiformis*	-
35.	*Echinocatcus grusonii*	Golden barrel/Golden bell
36.	*Echinocatcus ingens*	-
37.	*Echinocereus fitchii*	-
38.	*Echinocereus knippelianus*	-
39.	*Echinocereus pectinatus*	—
40.	*Echinocereus pentalophus*	-
41.	*Echinocereus purpureus*	-
42.	*Echinocereus triglochidiatus*	-
43.	*Echinopsis aurea*	-
44.	*Echinopsis calorubra*	-
45.	*Echinopsis eyriesii*	-
46.	*Echinopsis lau*	-
47.	*Echinopsis mammillosa*	-
48.	*Echinopsis multiplex*	-
49.	*Encephalocarpus strobiliformis*	-
50.	*Epiphyllum ackermannii*	-
51.	*Epithelantha micromeris*	-
52.	*Espostoa guentheri*	-
53.	*Espostoa lanta*	-
54.	*Espostoa melanostele*	-
55.	*Ferocactus alamosanus*	-
56.	*Ferocactus glaucescens*	-
57.	*Ferocactus horridus*	Fish-hook cactus
58.	*Ferocactus wislizenii*	-
59.	*Frailea castanea*	-
60.	*Gymnocalycium asterium*	-
61.	*Gymnocalycium denudatum*	Spider cactus
62.	*Gymnocalycium horstii*	-
63.	*Gymnocalycium mihanovichii*	Plain chin cactus
64.	*Gymnocalycium multiflorum*	-
65.	*Gymnocalycium quehlianum*	-
66.	*Gymnocalycium saglionis*	-

67.	*Haageocereus versicolor*	-
68.	*Hamatocatcus hamatacanthus*	-
69.	*Hamatocatcus setispinus*	-
70.	*Harissia eriophora*	-
71.	*Hatiora salicornioides*	-
72.	*Hildewintera aureispina*	-
73.	*Lemaireocereus pruinosus*	-
74.	*Leuchtenbergia principis*	Agave cactus
75.	*Lobivia famatimensis*	Orange cob cactus
76.	*Lobivia pentlandii*	-
77.	*Lobivia rubescens*	-
78.	*Lophocereus schottii*	-
79.	*Lephophora williamsii*	Peyote(or) whisky cactus
80.	*Lephophora ziegleri*	-
81.	*Machaerocereus eruca*	Creeping devil
82.	*Mammillaria baumii*	-
83.	*Mammillaria bocassana*	Powder puff
84.	*Mammillaria boolii*	-
85.	*Mammillaria candida*	Snow ball cactus
86.	*Mammillaria carmenae*	-
87.	*Mammillaria celsiana*	Showny pincushion
88.	*Mammillaria centricirrha*	-
89.	*Mammillaria compressa*	*Mother of hundred*
90.	*Mammillaria elegans*	-
91.	*Mammillaria elongata*	Golden stars/Gold lace cactus
92.	*Mammillaria geminispina*	-
93.	*Mammillaria glassii*	-
94.	*Mammillaria hahniana*	Old lady cactus
95.	*Mammillaria herrerae*	-
96.	*Mammillaria microcarpa*	-
97.	*Mammillaria plumosa*	Feather cactus
98.	*Mammillaria prolifera*	-
99.	*Mammillaria rodantha*	Rainbow pincushion
100.	*Mammillaria schiedeana*	-
101.	*Mammillaria spinosissima*	-

102.	*Mammillaria theresae*	-
103.	*Matucana paucicostata*	-
104.	*Matucana subgibosa*	-
105.	*Melocactus azureus*	-
106.	*Melocactus bahirensis*	Turk's head
107.	*Melocatcus broadwayii*	-
108.	*Melocactus matanzanus*	Melon cactus
109.	*Melocatcus saxicola*	-
110.	*Melocatcus schatzeli*	-
111.	*Melocatcus townsendii*	-
112.	*Monvillea spegazzinii*	-
113.	*Neogomesis agavoides*	-
114.	*Notocactus napina*	-
115.	*Notocactus graessnerii*	-
116.	*Notocactus haselbergii*	-
117.	*Notocactus leninghausii*	Golden ball
118.	*Notocactus magnificus*	-
119.	*Notocactus mammulosus*	Lemon ball
120.	*Notocactus scope*	Silver ball
121.	*Notocactus tabularis*	-
122.	*Obregonia denegrii*	Artichoke cactus
123.	*Opuntia clavariodes*	Sea oral
124.	*Opuntia cylindrica*	-
125.	*Opunita diademata*	-
126.	*Opunita falcata*	Tree opuntia
127.	*Opuntia ficus-indica*	Indian fig
128.	*Opuntia linguiformis*	-
129.	*Opuntia microdasys*	Bunny ears
130.	*Opuntia rufida*	Cinnamon cactus
131.	*Opuntia tunicate*	-
132.	*Opuntia vilis*	*Little tree cactus*
133.	*Oreocereus celsianus*	Old man of the Andes
134.	*Pachycereus pringlei*	-
135.	*Parodia chrysacanthion*	-
136.	*Parodia mairanana*	-

137.	*Peireskiopsis veluntina*	-
138.	*Pelecyphora asseliformis*	Hatchet cactus
139.	*Pereskia aculelata*	Barbados gooseberry/Lemon vine
140.	*Pereskia bleo*	Wax rose
141.	*Pereskia grandiflora*	-
142.	*Pseudopilocereus nobilis*	-
143.	*Rebutia heliosa*	-
144.	*Rebutia marsoneri*	-
145.	*Rebutia minuscula*	Red crown
146.	*Rebutia pulvinosa*	-
147.	*Rebutia senilis*	Fire crown
148.	*Rhipsalis burchellii*	Mistletoe
149.	*Schlumbergera bridgesii*	Christmas cactus
150.	*Schlumbergera russelliana*	-
151.	*Strombocactus disciformis*	-
152.	*Sulcorebutia arenacea*	-
153.	*Sulcorebutia flavissima*	-
154.	*Thelocactus bicolor*	*Glory of Texas*
155.	*Thelocactus hexaedrophorus*	-
156.	*Trichocereus candicans*	-
157.	*Trichocereus pasacana*	-
158.	*Trichocereus peruvianus*	-
159.	*Turbinicarpus lophophoroides*	-
160.	*Turbinicarpus schmiedickeanus*	-
161.	*Uebelamannai pectinifera*	-
162.	*Wilcoxia viperiana*	-
163.	*Zygocactus truncactus*	-

Chapter **11**

Flower Arrangements

I. General information

1. Japanese flower arrangement is called as *Ikebana*
2. Meaning for *Ikebana* is flowers kept alive
3. Western style of flower arrangement involves symmetrical arrangement of flowering plants in a vase
4. *Ikebana* style possess asymmetric flower arrangement
5. *Tatebana* is a type of flower arrangement in which flowers are placed in standing arrangement and are of tall nature
6. Basic lines of Japanese flower arrangement are heaven, man and earth
7. *'Shin'* in Japanese flower arrangement represents Heaven
8. *'Soe'* in Japanese flower arrangement represents Man
9. *'Hikae'* in Japanese flower arrangement represents Earth
10. Types of *Ikebaba* are *Moribana, Nageire, Jiyubana, Zaneika, Zeneibana* and *Morimono*
11. Ohara school which developed the style of flower arrangement in a shallow container is named as Moribana
12. Substyles of *Moribana* style are *Chokutai, Shatai* and *Suitai*
13. Meaning for *Chokutai* is upright
14. Meaning for *Shatai* is slanting
15. Meaning for *Suitai* is cascading
16. Flower arrangement by using piled - up flowers in a shallow vase is *Moribana*

17. Meaning of *Moribana* is piled-up flowers
18. Moribana is named as natural *Ikebana*
19. *Moribana* represents upright style arrangement
20. Vases used in *Moribana* style are of wide mouthed and often flat
21. Three basic lines of *Moribana* style are *Shin, Soe* and *Hikae*
22. *Shin* in *Moribana* style represents centre/ truth
23. *Soe* in *Moribana* style represents support
24. *Hikae* in *Moribana* style represents moderating/ overall balance
25. Three basic lines of *Moribana* style are collectively called as *Sansei*
26. Meaning for *Sansei* is harmony between nature and human
27. The term *'Utsuwa'* in *Ikebana* flower arrangement represents the shallow container of *Moribana*
28. Moribana style looks very natural and is referred as natural style of *Ikebana*
29. In *Moribana* style, flowers are not held by pinholders in the vase but are supported with the help of cross-bar fixture, single bar fixture and split bar fixture
30. For *Moribana* arrangement, heavy metal pin-holders are essential
31. In *Moribana* style, all the materials must be firmly fixed in the pin holder
32. In *Moribana* style, flower arrangement is viewed from the front side only
33. In *Moribana* style, half opened buds are used for flower arrangement
34. As per *Sogetsu* school, basic style of *Moribana* and *Nageire* arrangement is of two types
35. Two basic styles of *Moribana* and *Nageire* arrangements are basic upright and basic slanting
36. Basic upright style in *Moribana* arrangement consists of placing of pin holder at the frontal left corner of the container
37. Free style flower arrangement is named as *Rikka, Shoka, Nageire* and *Moribana*
38. The position of three basic lines in basic upright style are

Shin : At an angle of 10^0 to 15^0 on the left of the vertical 0^0 angle and bent forward by about 10^0

Soe : Left side of Shin at 45^0 angle from vertical line and caused to lean forward by 20^0 angle

Hikae : Right side of vertical line at an angle of 75^0 with a forward inclination of 30^0

39. In basic slanting style of *Moribana,* pin holder is placed at the rear of the container towards your right end
40. Flower arrangement in a tall vase is *Nageire*
41. In *Nageire* style, stems are usually inserted into vases without using *Kenzan*
42. Vases used in *Nageire* style are cylindrical, often tall and generally have a narrow mouth
43. The meaning for *Nageire* style is of slanted style
44. Three main branches of *Nageire* style are *Shin, Soe* and *Tai*
45. Sub styles of *Nageire* style are *Chokutai, Shatai* and *Suitai*
46. Height of *Shin* used in *Nageire* style is one and half of times the height of the vase plus width of the vase
47. Height of *Soe* used in *Nageire* style is two-third to three-quarters of *Shin* height in the vase
48. Height of *Tai* used in *Nageire* style is one-third to half of the *Shin* height in the vase
49. In case of *Nageire* arrangement, different types of bar or fixtures are used to support the flowering stem in the vase
50. Three types of fixtures used in *Nageire* arrangement are split-bar, single-bar and cross-bar
51. Bar-fixtures used in *Nageire* arrangement helps to support the short stem
52. In split-bar fixture, both the supporting bar and branch are given wedge shaped split and fitted with each other and is tied with a fine wire
53. In cross bar fixtures, two bars are crossed and fitted a little below the mouth of the vase

54. In Nageire, the branches are very often allowed to rest against the wall of the vase by using fixtures
55. In basic upright style in *Nageire,* cross bar fixture will divide the opening of vase into four quadrants and only one of such quadrant is used at a time for arranging the flowers
56. In basic upright style in *Nageire, Jushi* is placed in the same quadrant of three basic lines
57. In basic upright style in *Nageire, Shin* is fixed firmly in a split-bar fixture and cut end of shin should rest against the opposite inside wall of the vase
58. In basic upright style in *Nageire,* Soe is placed in other split-bar fixture and it can be held in position with the support of other stems
59. In basic slanting style of *Moribana, Shin* is placed at a relatively slanting angle of 45^0
60. Schools of *Ikebana* are *Ikenobo, Sogetsu* and *Ohara*
61. *Ikenobo* school is located in Rokkakudo temple in Kyoto
62. The founder of *Ikenobo* school is Buddhist priest Ikenobo Senkei
63. The founder of Sogetsu school is Teshigahara Sofu
64. *Sogetsu* school which developed the modern style is named as *Zeneibana*
65. *Zeneibana* style involves using new materials such as steel and plastic without using plants
66. The first modern school of *Ikebana* was established in Japan is Ohara
67. Ohara school consider the western style while developing flower arrangement techniques
68. Free style of *Ikebana* is called as *Jiyuka*
69. *Jiyuka* style was developed by Ohara
70. *Rikka* style of flower arrangement involves standing flowers
71. *Shoka* style of flower arrangement involves living flowers
72. *Heika* style incorporates the styles of *Rikka, Shoka* and *Seika*
73. *Rikka* style of *Ikebana* adopt strict rules
74. *Shoka* and *Seika* of *Ikebana* adopt less rigid rules

75. Two varieties of *Rikka* style are *Rikka Shofutai* and *Rikka Shimputai*
76. *Rikka* style act as a foundation for *Shoka, Moribana* and *Nageire* styles
77. *Rikka* style of flower arrangement was established during Muromachi period
78. In Japanese,the word *Kenzan* used in flower arrangement represents Pinholder
79. The height of vase used in Rikka is 20-30cm
80. *Rikka Shofutai* involves wiring technique
81. The most formal style of *Ikebana* which express the beauty of natural landscape is *Rikka Shofutai*
82. Nine main stems used in *Rikka Shofutai* style are Shin, Shoshin, Soe, Uke, Nagashi, Mikoshi, Hikae, Do, Maeoki
83. Nine main stems used in *Rikka Shofutai* style are collectively called as *Yakueda*
84. *Rikka Shofutai* style possess fixed point of departure is called as *De* in Japanese
85. The term *Mizugiwa* used in *Rikka Shofutai* style represents the stem which are of clean, single, vertical line raising from the centre of the Kenzan
86. The main parts of *Rikka Shimputai* style are *Shun* and *Yo*
87. *Rikka Shimputai* style does not involve fixed point of departure
88. *Rikka Shimputai* style give more relaxed feeling and fresh appearance
89. In the large vase of *Ikebana* flower arrangement, *Shin* lies at twice the measurement of the diameter and height of the vase {2 x (diameter +height of the vase)}
90. In the large vase of *Ikebana* flower arrangement, *Soe* lies at the $3/4^{th}$ length of *Shin* which is positioned at 1.5 times height of the vase
91. In the large vase of *Ikebana* flower arrangement, *Hikae* lies at 1/2 of the length of *Soe*
92. A simple diagram which helps to place branches and spikes in the metal holder of *Ikebana* arrangement is *Kakeizu*
93. The term *Kakeizu* in flower arrangement shows a frontal and overhead view

94. Modern *Ikebana* style is *Hanaisho*
95. *Morimono* is named as English flower arrangement
96. *Morimono* style comprises of fruits, vegetables and flowers
97. Most difficult *Ikebana* style is *Shoka*
98. Two types of *Shoka* style are *Shoka Shofutai* and *Shoka Shimputai*
99. The vase used for *Shoka* style is symmetrical and open at the top
100. *Shoka Shofutai* style consists of *Shin, Soe* and *Tai*
101. *Shin* used in *Shoka Shofutai* style is 2-3 times the height of the vase
102. *Soe* used in *Shoka Shofutai* style is 2/3rd the height of the *Shin*
103. *Tai* used in *Shoka Shofutai* style is 1/3rd the height of the *Shin*
104. *Shoka Shimputai* express well ordered beauty in which *Shin: Soe: Tai* ration is 7:5:3
105. *Shoka Shimputai* emphasis unbalanced beauty
106. Characteristics of *Shoka Shimputai* are brightness and sharpness
107. *Shoka Shimputai* consists of two parts *Shu* and *Yo*
108. The term '*Kabin*' in *Ikebana* flower arrangement represents the tall used in *Heika* style
109. The lazy 'S' arrangement used in artificial and dried flower arrangement is named as **Hogarth's curve**
110. *Jiyubana* style comprises of wood, metal or any other material used with flowers
111. *Jiyubana* style can be arranged in both *Moribana* and *Nageire* style
112. *Zeneika* style can be arranged with straight material at an uneven height
113. *Zeneika* style does not simulate nature
114. In *Zeneibana* style beautiful sculpture is created using wood, stone, rocks and metal
115. *Zeneibana* style simulates the nature
116. 'Heaven' is placed in flower vase at an angle of 45^0C
117. Japanese flower arrangement is based on spiritual themes

118. English flower arrangement is based on artistic themes

119. English flower arrangement comprises of fruits, flower and foliages which imparts collective mass effect

120. In Western style of flower arrangement, height of flower is fixed at 2 times of flower vase height

121. Japanese flower arrangement was started by Buddhist monk Semmu in 621 A.D

122. Morimono style of *Ikebana* resembles English flower arrangement

123. Meaning for *Ikenobo* is the arrangement of flowers of the hermitage along a pond

124. *Zeneika* style is named as Abstract style

125. In *Zeneika* style, plant materials are used in combination with inert material

126. In *Zeneika* style, the arrangement will be pleasing and in agreement with the surroundings of the room

127. *Zeneibana* style is named as Avant-garde

128. *Zeneibana* style is depicting seaside or any other natural scenery

129. In any flower arrangement, the *Shin* should always be the tallest stem, followed in terms of height by *Soe* and *Hikae*

130. *Jushi* stems are called as fillers

131. In a standard size or small size vase, the length of *Shin* will be one and a half times the height of the vase

132. *Soe* is 3/4th of *Shin*

133. *Hikae is 3/4th of Soe*

134. For *Morimono* style, vase will be of wooden material, tray or basket or a shallow bowl

135. If *Morimono* arrangement is made on a wooden base, it is called as '*Dai*' in Japanese

136. In *Morimono* arrangement, the importance given for flower is secondary

137. Fruits used in *Morimono* arrangement are grapes, orange, grapefruit, pineapple and apple

138. Vegetables used in *Morimono* arrangement are red tomatoes, capsicum, brinjal and carrots

139. In *Morimono* arrangement, large bulky and very tall flowers should not be used

140. For foliage arrangement, tall leaves such as palms, cycads, sansevierias, lilies, screwpine can be combined with dwarf oval colourful foliage such as coleus, rex begonia

141. For cascade arrangement using foliage, leaves such as *Pilea muscosa, Zebrina pendula, Hedera helix,* Crotons, *Magnolia grandiflora* are used

142. Rangoli is the drawing of pattern with the help of dry colour or a coloured paste on the floor

143. Floral bouquets are arrangements made with a group of cut flowers and is an usual practice to present a nice floral bouquet on the occasion of birthdays, marriages or to welcome important dignitaries

144. Buttonholes is a fascinating item worn by males in their coat collars near the chest

145. In West Bengal, Goddess Kali is offered with garlands of red hibiscus

146. In West Bengal, during Annaprasana, floral crown made up of tuberose is used

II. Fill in the blanks

1. Japanese flower arrangement is based on.................
2. The founder of Sogetsu school is.................
3. *Ikenobo* school is located in.................
4. *Rikka* style of *Ikebana* adopt.................
5. The height of vase used in Rikka is.................
6. Meaning for *Chokutai* is.................
7. Meaning for *Shatai* is.................
8. Meaning for *Suitai* is.................
9. The lazy 'S' arrangement used in artificial and dried flower arrangements is named as.................
10. *Zeneika* style does not simulate.................

III. Choose the best answer

1. In flower arrangement, heaven is placed at an angle of .... in flower vase

   15^0 30^0 45^0

2. *Jushi* stems are called as.................

   Flower Fillers Pin holer

3. Flower arrangement by using piled flowers in a shallow vase is.................

   Moribana *Morimono* *Zeneibana*

4. *Morimono* style comprises of ....

   Fruits Vegetables Fruits, Vegetables and flowers

5. Flower arrangement in a tall vase is.................

   Nageire *Moribana* *Morimono*

6. School of Ikebana is.................

   Zeneibana *Hanaisho*

7. Style of flower arrangement which is named as natural *Ikebana* is.................

   Morimono *Zeneibana*

8. Modern *Ikebana* style is.................

   Zeneibana *Hanaisho* *Shoka*

9. Most difficult *Ikebana* style is.................

   Zeneibana *Hanaisho* *Shoka*

10. Type of flowers used in *Moribana* style is.................

    Bud stage Fully opened flower

IV. Match the following

1. *Shin* : Earth
2. *Soe* : Heaven
3. *Hikae* : Man
4. *Kenzan* : Upright style

5.	*Nageire*	:	Avant-garde
6.	*Moribana*	:	Coat collars
7.	Buttonholes	:	Pinholder
8.	*Zeneibana* style	:	Abstract style
9.	*Zeneika*	:	Unbalanced beauty
10.	*Shoka Shimputai*	:	Slanted style

Answers

II. Fill in the blanks

1. Spiritual themes
2. Teshigahara sofu
3. Kyoto
4. Strict rules
5. 20-30cm
6. Upright
7. Slanting
8. Cascading
9. Hogarth's curve
10. Nature

III. Choose the best answer

1. 45^0
2. Fillers
3. *Moribana*
4. Fruits, vegetables and flowers
5. *Nageire*
6. *Sogetsu*
7. *Moribana*
8. *Hanaisho*
9. *Shoka*
10. Half opened buds

IV. Match the following

1. Heaven
2. Man
3. Earth
4. Pinholder
5. Slanting style
6. Upright style
7. Coat collars
8. Avant-garde
9. Abstract style
10. Unbalanced beauty

Chapter **12**

Bonsai

I. General information

1. Bonsai is defined as Japanese art of growing trees & shrubs by extreme dwarfing in a shallow container
2. Bonsai, the name is derived from two words, Bon & Sai
3. The meaning for the word 'Bon' is **tray or shallow container**
4. The meaning for the 'Sai' is **to grow**
5. Origin of Bonsai is **china**
6. In Mamie bonsai, plant height is **2-6 inches**
7. In small bonsai, plant height is **6-12 inches**
8. In medium bonsai, plant height is **12-24 inches**
9. In large bonsai, plant height is **24-60 inches**
10. The ideal time for planting bonsai is **February- March, July- August**
11. Different styles of bonsai are Formal upright, Informal upright, Slanting trunk, Semi cascade, Cascade or hanging, broom, windswept, Twisted, Drift-wood, Weeping branch, split-trunk, Bunjin or Literati & Exposed –root style, Multiple-trunk, Twin or triple trunk, Clump style, raft style
12. Two forms of Chinese bonsai are **Penjai & Penjing**
13. The meaning for the word 'Penjai' is **a tree planted in a container without any landscape**
14. The meaning for the word 'Penjing' is **a tree planted in a container or on a tray with a landscape**
15. Two forms of Raft style bonsai are **Straight raft, Sinuous raft**

16. Tree species which are suitable for broom style bonsai are ***Thuja orientalis, Malpighia coccigera***
17. Tree species which are suitable for bunjin style bonsai are ***Callistemon lanceolatus,* Casuarina**
18. Tree species which are suitable for windswept style bonsai are **Juniper, Casuarina, etc.,**
19. Tree species which are suitable for split-trunk style bonsai are **Juniper, Ceder, etc.,**
20. Tree species which are suitable for exposed-root bonsai are ***Ficus bengalensis***
21. Tree species which are suitable for making bonsai of natural group or forest style is **Juniper, Casuarina, etc.,**
22. Pots of different shapes used for bonsai culture are square, round, oval, rectangular, long, narrow, hexagonal, octagonal
23. For upright style bonsai, pots of **square or rectangular** shape are preferred
24. For informal style bonsai, pots **of round or oval** shape are preferred
25. For cascade style bonsai, **deep pots** are preferred
26. For group style bonsai, pots of **oval shape** is preferred
27. Pot colour of bonsai culture suitable for **evergreen plants** is **brown to brownish red, black, grey colours**
28. Pot colour of bonsai culture suitable for deciduous trees is **green, blue or white**
29. Pot colour of bonsai culture suitable for flowering trees is **earthy brown- coloured pots**
30. Pot colour of bonsai culture suitable for fruiting trees like citrus, pomegranate, ber is **colourful pots**
31. Pot colour of bonsai culture suitable for red-fruited trees is **white**
32. Pot colour of bonsai culture suitable for green fruited trees is **blue**
33. Type of wire which is most suitable for making bonsai is **copper**
34. Thickness of copper wire used for bonsai culture **is 1.65 mm-2.76 mm (12-18 guage)**
35. In round or square shaped pot, positioning of plants for bonsai culture is in the **centre**

36. In oval or rectangular pot, positioning of plants for bonsai culture is at **one-third of length of pot from any end**
37. Ideal time for repotting in bonsai culture is **early spring or monsoon**
38. Duration taken for repotting in bonsai culture for evergreen trees like pine & juniper is **3-5 years interval**
39. Duration taken for repotting in bonsai culture for flowering & fruiting trees is in **one year** interval
40. Formal upright style of bonsai is called as **Chokkan**
41. Informal upright style of bonsai is called as **Moyogi**
42. Slanted style of bonsai is called as **Shakan**
43. Cascade-style of bonsai is called as **Kengai style**
44. Semi- style of bonsai is called as Han **Kengai style**
45. Raft- style of bonsai is called as **Netsuranari**
46. Literati - style of bonsai is called as **Bunjin-gi style**
47. Group or forest - style of bonsai is called as **Yose Ue**
48. Broom - style of bonsai is called as **Hokidachi**
49. Multi-trunk style of bonsai is called as **Ikadabuki**
50. Shari style of bonsai is called as **Sharimiki**
51. Root- over rock style of bonsai is called as **Sekijoju**
52. Growing in a rock style of bonsai is called as **Ishizuke**
53. Trees suitable for making bonsai are

    1) Ficus bengalensis – Banyan tree

    Ficus benjamina – Java fig, weeping fig

    Ficus religiosa – Peepal tree

    Ficus nitida – Indian laurel

    Ficus triangularis

    Ficus diversofolia

    Ficus infectoria

    Ficus rumphii

    Ficus microcarpa – Chinese banyan fig

Ficus neriifolia – Willow-leaved fig

Ficus rubiginosa – Port Jackson fig

2) *Callistemon lanceolatus* – Red Bottle Brush

3) *Diospyros peregrine* – River ebony

4) Cordia sebestena

5) *Casuarina equisetifolia* – Casuarina

6) *Juniperus chinensis* – Chinese Juniper

7) *Juniperus horizontalis* – Creeping Juniper

8) Malphigia coccigera

9) *Malphigia glabra* – Barbados cherry

10) *Punica granatum* – Pomegranate

11) *Pinus roxburghii* – Indian pine, chir

12) *Salix babylonica* – Weeping willow

13) *Pinus clausa* – Sand pine

14) *Pinus mugo* – Mugo pine, Mountain pine

15) *Pinus parviflora*- Japanese white pine

16) *Pinus thunbergii* – Japanese black pine

17) *Pinus virginiana* – Virginia pine

18) *Pinus ponderosa* – Western yellow pine

54. Climbers suitable for Bonsai making is **Petrea &*Hedera helix***

55. Fruit trees suitable for Bonsai making are **Pomegranate, Ber, Mango, Sapota, etc**

56. Shrubs suitable for Bonsai culture are **Bougainvillea, Lagerstroemia, *Jasminum nudiflorum, Hibiscus syriacus*, Ligustrum, *Murraya paniculata***

57. Windswept style is fukinagashi

58. Driftwood style is sharimiki

59. Clinging to rock style is hitsuki

60. Bonsai plants propagated by seeds are *Adansonia digitata, Azadirachta indica, Bassiamonosperma, Cassia fistula, Commiphora mukul, Diospyros*

exsculpta, Ficus religiosa, Lagerstroemia speciosa, Lagerstroemia thorelli, Millettia ovalifolia, Terminalia arjuna, Terminalia bellerica, Anacardium occidentale, Bauhinia purpurea, Ficus glomerata, Gardenia latifolia, Grevillea robusta, Michelia alba

61. Shrubs suitable for making bonsai and are propagated by cuttings and air-layering are *Adenium obesum, Caesalpinia pulcherrima, Calliandra* spp, *Duranta repens, Hibiscusrosa-sinensis, Jatropha* sp, *Tibouchina semidecandra, Kopsia fruiticosa, Lantana camara, Lawsonia alba, Malpighia coccigera, Rosa* sp, *Murraya paniculata, Murraya koenigii*

62. Bonsai plants grown for flowers are *Albizia lebbeck, Bauhinia purpurea, Bauhinia variegata, Cassia nodosa, Delonix regia, Gardenia gummifera, Gardenia latifolia, Gardenia variegata, Holarrhena antidysenteria, Ixora parviflora, Lawsonia alba, Michelia champaca, Mitragyna parviflora, Ravenia spectabilis*

63. Fruit bearing bonsai plants are *Annona squamosa, Achras sapota, Averrhoa carambola, Carissa carandas, Citrus acida, Citrus aurantium, Citrus aurantifolia, Citrus jambheri, Citrus maxima, Citrus reticulata, Citrus paradisi, Citrus sinensis, Fortunella japonica, Limonia trifolia, Morus indica, Psidium cattleyanum, Syzygium cumini, Ziziphus mauritiana*

64. Straggling shrubs (climbers) suitable for making bonsai by pruning are *Allamanda cathartica, Artabotrys odoratissimus,* Bougainvillea, *Petrea volubilis, Tecomaria capensis*

65. Evergreen shrubs & tress suitable for bonsai making are *Alstonia scholaris, Araucaria cunninghamii, Araucaria cookii, Casuarinas equisetifolia, Crassula arborescens, Cryptomeria japonica, Ficus retusa, Carmona microphylla, Juniperus chinensis, Limonia acidissima, Pandanus andamanensium, Mimusops elengi, Mimusops hexandra, Pinus longifolia, Portulacaria afra, Thevetia nerifolia, Thuja orientalis, Ficus bengalensis, Ficus elastica, Ficus nitida, Ficus rumphi*

66. Light intensity required for bonsai making in indoor condition is 200-1000 lux

II. Fill in the blanks

1. Japanese art of growing trees & shrubs by extreme dwarfing is..............
2. Origin of Bonsai is..............
3. In Mamie bonsai, plant height is..............
4. In small sized bonsai, plant height is..............

5. In medium sized bonsai, plant height is..............
6. Tree species which is suitable for exposed-root style bonsai is..............
7. Two forms of Chinese bonsai are..............
8. Two forms of Raft style bonsai are..............
9. The shape of pot suitable for cascade style bonsai is..............
10. The shape of pot suitable for group style bonsai is..............
11. Pot colour of bonsai culture suitable for red-fruited trees is..............
12. Pot colour of bonsai culture suitable for green fruited trees is..............
13. Type of wire which is most suitable for making bonsai culture is..............
14. Root- over rock style of bonsai is called as..............
15. Light intensity required for bonsai making in indoor condition is..............

III. Choose the best answer

1. Straggling shrub (climber) which is suitable for making bonsai by pruning is..............

   Allamanda cathartica *Hibiscus mutabilis* *Caesalpinia pulcherrima*

2. Driftwood style of bonsai is ....

   Sharimiki Hitsuki Fukinagashi

3. Cascade-style of bonsai is called as ....

   Hitsuki Fukinagashi Kengai

4. Pot colour of bonsai culture suitable for flowering trees is ....

   Green Blue Earthy brown- coloured pots

5. Tree species which is suitable for broom style bonsai is ....

   Citrus aurantifolia *Thuja orientalis* *Averrhoa carambola*

6. Climber which is suitable for Bonsai making is ....

   Campsis grandiflora *Hedera helix* *Ficus repens*

7. Shrub which is suitable for Bonsai culture is ....

   Murraya paniculata *Buddleia asiatica* *Gardenia jasminoides*

8. Bonsai plant which is grown for flower is ....

   Peltophorum ferrugineum *Delonix regia* *Spathodea campanulata*

9. Bonsai plant which is propagated by seed is ....

   Azadirachta indica *Adenium obesum* *Caesalpinia pulcherrima*

IV. Match the following

1.	Cascade-style of bonsai	:	*Callistemon lanceolatus*
2.	Penjai	:	Juniper
3.	Bon	:	Moyogi
4.	Bunjin style bonsai	:	Kengai style
5.	Split-trunk style bonsai	:	Tree planted in a container without any landscape
6.	Informal upright style of bonsai	:	Tray or shallow container
7.	Group or forest style of bonsai	:	Fukinagashi
8.	Clinging to rock style	:	Yose Ue
9.	Pomegranate	:	hitsuki
10.	Windswept style	:	Colourful pot

Answers

II. Fill in the blanks

1. Bonsai	2. China
3. 2-6 inches	4. 6-12 inches
5. 12-24 inches	6. *Ficus bengalensis*
7. Penjai & penjing	8. Straight raft, sinuous raft
9. Deep pot	10. Oval shape

11. White
12. Blue
13. Copper
14. Sekijoju
15. 200-1000 lux

III. Choose the best answer

1. *Allamanda cathartica*
2. Sharimiki
3. Kengai style
4. Earthy brown- coloured pots
5. Thuja orientalis
6. Hedera helix
7. *Murraya paniculata*
8. *Delonix regia*
9. *Azadirachta indica*

IV. Match the following

1. Kengai style
2. Tree planted in a container without any landscape
3. Tray or shallow container
4. *Callistemon lanceolatus*
5. Juniper
6. Moyogi
7. Yose Ue
8. hitsuki
9. Fukinagashi
10. Colourful pot

Chapter **13**

Turf Grasses

I. General information

1. Cool season turf grasses are **Perennial rye grass, Creeping bent grass, Colonial bent grass, Kentucky blue grass, Canada blue grass, *Poa* spp, *Festuca* spp, *Agrostis* spp, *Lolium* spp**
2. Warm season turf grasses (25-35°C) are ***Cynodon* sp, *Zoysia* sp, *Stenotaphrum secundatum, Paspalum notatum, Eremochloa ophiuroides, Axonopus* sp, *Pennisetum clandestinum, Buchloe dactyloides***
3. Broad leafed turf grasses are *Arachis pintoi,* ***Dichondra repens***
4. Based on mowing height, turf grasses are classified into **tall, medium & dwarf**
5. Turf grasses which needs high mowing height of >3.5 cm are **Tall fescue, sheep fescue, Canada blue grass, fine fescue, wheat grass, orchard grass**
6. Turf grasses which requires medium mowing height of d" 3.5 to < 1.0 cm are **perennial rye grass, Kentucky blue grass, Buffalo grass, etc.,**
7. Turf grasses which requires low mowing height of d" 1.0 cm are **colonial bent grass, creeping bent grass, Annual blue grass**
8. Turf grasses are classified based on their recuperative capacity into **fast, medium & slow**
9. Turf grasses of fast recuperative ability are **creeping bent grass, Kentucky blue grass, Bermuda grass, Perennial rye grass, Fine fescue, Timothy**
10. Turf grasses of medium recuperative ability are **colonial bent grass, velvet bent grass, sheep fescue**

11. Turf grasses of slow recuperative ability are **Buffalo grass, wheat grass, orchard grass**
12. Turf grasses which are classified based on wear resistance are **high, moderate & low**
13. Turf grasses which exhibit highly wear resistant are **tall fescue, perennial rye grass, Canada blue grass, sheep fescue**
14. Turf grasses of moderately wear resistant are **fine fescue, buffalo grass, Kentucky blue grass, Timothy, etc.,**
15. Turf grasses of low wear resistant are **creeping bent, colonial bent grass & velvet bent grass**
16. Turf grasses which are classified based on thatching tendency are **high, medium & low**
17. Turf grasses with high thatch forming tendency (>2 cm) are **creeping bent grass, velvet bent grass, colonial bent grass, Kentucky blue grass, fine fescue**
18. Turf grasses with medium thatch forming tendency (1-2 cm) are **perennial rye grass, tall fescue, sheep fescue, Timothy**
19. Turf grasses with low thatch forming tendency (<1 cm) are **Alkali grass, Buffalo grass, wheat grass, orchard grass**
20. On the basis of shoot density, turf grasses are classified into **high, medium & low**
21. Turf grasses with **high shoot density** are **creeping bent grass, colonial bent grass, sheep fescue, fine fescue**
22. Turf grasses of **medium shoot density** are **Kentucky blue grass, Perennial rye grass, Buffalo grass, tall fescue, Alkali grass**
23. Turf grasses of **low shoot density** are **Canada blue grass, wheat grass, orchard grass**
24. Turf grasses which responds well for mowing are **Kentucky blue grass, Bermuda grass, colonial bent grass, creeping bent grass**
25. Turf grasses which responds poor for mowing are **Tall fescue, sheep fescue, perennial rye grass**
26. Turf grasses which **tolerate high shade** are **Rough stalk blue grass, Fine fescue, Sheep fescue, Velvet bent grass, Orchard grass**
27. Turf grasses which **tolerate medium shade** are **colonial bent grass, Tall fescue, Canada blue grass, Annual blue grass, creeping bent grass**

28. Turf grasses which **tolerate low shade** are **Kentucky blue grass, perennial rye grass, Buffalo grass**

29. Turf grasses which **tolerate high salt content** are **Alkali grass, Buffalo grass, wheat grass, Bermuda grass, Zoysia grass, St. Augustine grass**

30. Turf grasses which **tolerate moderate salt content** are **creeping bent grass, sheep fescue, tall fescue, Timothy, Perennial rye grass, Canada blue grass, fine fescue**

31. Turf grasses which **tolerate high soil acidity** of (4.0 < 5.5 pH) are **velvet bent grass, tall fescue, fine fescue, sheep fescue, centipede grass**

32. Turf grasses which **tolerate medium soil acidity** of (5.5-6.8 pH) are **colonial bent grass, creeping bent grass, orchard grass, perennial rye grass, Kentucky blue grass**

33. Turf grasses which **tolerate low soil acidity** of (>6.8 pH) are **wheat grass, Alkali grass, Timothy**

34. Turf grasses which possess high flooding tolerance are **Rough stalk blue grass, creeping bent grass, tall fescue, Alkali grass, Colonial bent grass, Velvet bent grass, Bermuda grass, Bahia grass**

35. Turf grasses which possess medium flooding tolerance are **perennial rye grass, Timothy, orchard grass, fine fescue**

36. Turf grasses which possess low flooding tolerance are **Kentucky blue grass, wheat grass, sheep fescue, Buffalo grass, Canada blue grass**

37. Turf grasses which possess high drought tolerance are **Buffalo grass, crested wheat grass, western wheat grass, sheep fescue, Canada blue grass**

38. Turf grasses which possess medium drought tolerance are **fine fescue, tall fescue, orchard grass, Kentucky blue grass, perennial rye grass**

39. Turf grasses which are of low drought tolerance are **Annual blue grass, colonial bent grass, creeping bent grass, timothy, rough stalk blue grass**

40. Turf grasses which tolerate high temperature (>35°C) are **Buffalo grass, tall fescue, wheat grass, sheep fescue, Bermuda grass**

41. Turf grasses which tolerate medium temperature (>15°C- 35°C) are **creeping bent grass, Alkali grass, sheep fescue, Canada blue grass, Kentucky blue grass, colonial bent grass, fine fescue**

42. Turf grasses which tolerate low temperature (<15°C) are **perennial rye grass, orchard grass, timothy**
43. Turf grasses of high cold tolerance are **creeping bent grass, perennial rye grass, orchard grass, Kentucky blue grass, timothy & colonial blue grass**
44. Turf grasses of medium cold tolerance are **fine fescue, alkali grass, tall fescue, perennial rye grass**
45. Turf grasses with less low cold tolerance are **Buffalo grass, Bermuda grass, wheat grass**
46. Turf grasses which are highly susceptible to diseases are **creeping bent grass, colonial bent grass, velvet bent grass**
47. Turf grasses which are less susceptible to diseases are **Kentucky blue grass, fine fescue, perennial rye grass, tall fescue**
48. Turf grasses which need heavy fertilizers are **creeping bent grass, perennial rye grass, velvet bent grass**
49. Turf grasses which need moderate fertilizers are **annual blue grass, Canada blue grass, buffalo grass, tall grass, fine fescue, etc.,**
50. Turf grasses which need less fertilizers are **orchard grass, sheep fescue, wheat grass**
51. Turf grasses which possess **coarse leaf texture** of (> 5 mm leaf width) are **orchard grass, Tall fescue & wheat grass**
52. Turf grasses which possess **fine leaf texture** of (<2 mm leaf width) are **creeping bent grass, fine fescue, sheep fescue, velvet bent grass**
53. Turf grasses which possess **medium coarse leaf texture** of (2.0 -5 mm leaf width) are **Kentucky blue grass, perennial rye grass, Canada blue grass, Timothy**
54. Turf grasses which possess **medium fine grasses** are **buffalo grass, colonial bent grass & alkali grass**
55. Turf grasses with fast establishment rate are **perennial rye grass, tall fescue, fine fescue, creeping bent grass, colonial bent grass**
56. Turf grasses with medium establishment rate are **Kentucky blue grass, Timothy, velvet bent grass**
57. Turf grasses with slow establishment rate are **wheat grass, buffalo grass, sheep fescue**
58. Major diseases of turf grasses are **dollar spot & fairy ring**

59. Annual monocot weeds found in lawns are **large hairy crab grass, smooth crab grass, annual blue grass**

60. Annual dicot weeds found in lawns are **black medic, purslane, knot weed, prostrate spurge**

61. Perennial monocot weeds found in lawns are **Quack grass, orchard grass, purple nutsedge, Bermuda grass, yellow wood sorrel, white clover, black horn plantain, Dandelion, Thistles, Chicory**

62. Plant growth retardants used in Bahia grass, Bermuda grass & tall fescue are **Chlorsulfuron, Sulfometuron- methyl, Metsulfuron, Sethoxydin, Maleic hydrazide, Glyphosate, Imazapic, Mefluidide**

63. Plant growth retardants used in Bermuda grass, tall fescue & St. Augustine grass are **Flurprimidol, Trinexapac- ethyl, Paclobutrazol**

64. Seed rate to establish lawn for 100 m^2 is **250-300 g**

65. In turf grass establishment, the process of spreading stolons on the soil surface is known as **stolonizing**

66. Stolonizing is also called as **Dibbling & Stolonization**

67. Expensive method of planting turf grasses is **sodding**

68. The grass height in which first mowing should be done in a newly established lawn is **5-6 cm**

69. Dethatching is a process in which **removing** a layer of dead & decaying organic debris and living roots crown & stems of the grass from the top layer

70. Grasses suitable for tennis court are **Bermuda, perennial rye & creeping bent grass**

71. Grasses suitable for athletic fields are **Kentucky blue grass, Bermuda grass, perennial rye grass, tall fescue grass**

72. Grasses suitable for cricket pitches and out fields are **Bermuda grass and rye grass**

73. In Golf course, turf grasses suitable for "**greens**" are **annual blue grass, creeping bent grass, hybrid Bermuda grass, Zoysia grass & seashore paspalum**

74. In golf course, turf grasses suitable for "**fairways**" are cultivars of Kentucky blue grass (such as Absolute, Award, Bluemoon, Nuglade, Rambo, Unique), Bermuda grass, Buffalo grass

75. In golf course, turf grasses suitable for "**rough area**" are **Kentucky blue grass, buffalo grass, kikuyu grass**

76. In golf course, turf grasses suitable for **"Tees"** are *creeping bent grass, perennial rye grass, Bermuda grass, Zoysia grass, buffalo grass, seashore paspalum*

77. **Tifgreen Bermuda grass** is a hybrid of Cynodon dactylon and C. transvaalensis. This hybrid has the capacity to resists disease, poor irrigation circumstances, high traffic and droughts. It is low growing, dark green grass and spreads quickly. This grass can regenerate quickly as well. It is for this reason that it's quite extensively used in Golf course greens, tees, lawns and commercial landscapes. Weed control chemicals do not damage Tifgreen grass.

78. **Tifdwarf** (*C. dactylon* x *C. transvaalensis*), this variety is a vegetative mutant of Tifgreen. Tifdwarf is mostly similar with Tifgreen but its leaves and internodes are much shorter than Tifgreen. The green color is also darker than Tifgreen. Tifdwarf becomes reddish-purple in color immediately after the winter. For its superior putting quality it is popularly used for golf greens, tennis courts and bowling greens and such.

79. The common name of following species of turf grasses are

    Agrostis canina – velvet bent grass

    Agrostis capillaries – colonial bent grass

    Agrostis castellana – Dryland bent grass

    Agrostis giganteaca – Red top

    Agrostis spp – Bent grass

    Agrostis stolonifera – creeping bent grass

    Buchloe dactyloides - buffalo grass

    Cynodon sp – Bermuda grass

    Eremochloa ophiuroides – centipede grass

    Festuca arundinaceae – Tall fescue

    Festuca ovina – Sheep fescue

    Festuca rubra – creeping red fescue

    Festuca sp – Fescue grass

    Festuca tryachyphylla – Hard fescue

    Lolium multiflorum – Annual rye grass

Lolium perenne – Perennial rye grass

Lolium sp – Rye top

Paspalum notatum – Bahia grass

Paspalum vaginatum – seashore paspalum

Pennisetum clandestinum – kikuyu grass

Poa alpine – Alpine blue grass

Poa annua – Annual blue grass

Poa arachnifera – Texas blue grass

Poa bulbosa – Bulbous blue grass

Poa compressa – Canada blue grass

Poa pratensis - Kentucky blue grass

Poa sp. – Blue grasses

Poa supina – Supina blue grass

Poa trivalis – Rough blue grass

Stenotaphrum secundatum – St. Augustine grass

Zoysia sp – Zoysia grass/Korean grass/Mexican grass

II. Fill in the blanks

1. Seed rate to establish lawn for 100 m^2 is.................
2. Expensive method of planting turf grasses is.................
3. The grass height at which first mowing should be done in a newly established lawn.................
4. In turf grass establishment, the process of spreading stolons on the soil surface is called as.................
5. A process which helps to remove a layer of dead & decaying organic debris and living roots crown & stems of the grass from the top layer is called as.................

III. Choose the best answer

1. Turf grass which needs high mowing height of >3.5 cm is.................

   Colonial bent grass Creeping bent grass Tall fescue grass

2. Broad leafed turf grass is.................

   Dichondra repens   Bermuda grass   Perennial rye grass

3. Turf grass of fast recuperative ability is.................

   Buffalo grass   Bermuda grass   orchard grass

4. Turf grass which exhibit higher wear resistance is.................

   Sheep fescue   Creeping bent   Colonial bent grass

5. Turf grass which possesses high thatch forming ability (>2 cm) is.................

   Buffalo grass   Wheat grass   Fine fescue

6. Turf grass with high shoot density is.................

   Canada blue grass   Wheat grass   Creeping bent grass

7. Turf grass which responds well for mowing is.................

   Tall fescue   Sheep fescue   Kentucky blue grass

8. Turf grass which tolerates high shade is.................

   Kentucky blue grass   Perennial rye grass   Fine fescue

9. Turf grass which tolerates high salt content is.................

   Creeping bent grass   Sheep fescue   Zoysia grass

10. Turf grass which tolerates high soil acidity of (4.0 < 5.5 pH) is.................

    Wheat grass   Alkali grass   Centipede grass

11. Turf grass which possess high flooding tolerance is

    Sheep fescue   Buffalo grass   Bermuda grass

12. Turf grass which possesses high drought tolerance is.................

    Colonial bent grass   Creeping bent grass   Buffalo grass

13. Turf grass which tolerates high temperature (>35°C) is.................

    Perennial rye grass   Timothy grass   Tall fescue

14. Turf grass of high cold tolerance is.................

    Buffalo grass   Kentucky blue grass   Bermuda grass

15. Turf grass which is highly susceptible to disease is.................

Perennial rye grass | Creeping bent grass | Tall fescue

16. Turf grass which needs heavy fertilization is.................

    Velvet bent grass | Buffalo grass | Bermuda grass

17. Turf grass which possesses coarse leaf texture (> 5 mm leaf width) is.................

    Fine fescue | Sheep fescue | Tall fescue

18. Turf grass which possesses fine leaf texture (< 2 mm leaf width) is.................

    Fine fescue | Tall fescue | Wheat grass

19. Turf grass with fast establishment rate is.................

    Buffalo grass | Sheep fescue | Tall fescue

20. Major disease of turf grasses is.................

    Dollar spot | Pythium | Phytophthora

IV. Match the following

1. Tennis court	Dandelion
2. Athletic fields	Bent grass
3. Cricket pitches	Perennial rye grass
4. Perennial monocot weed	Creeping bent grass
5. *Agrostis* spp	Tall fescue grass
6. *Lolium perenne*	Rye grass
7. *Buchloe dactyloides*	Sheep fescue
8. Warm season turf grass	*Stenotaphrum secundatum*
9. Cool season turf grass	*Agrostis* spp.
10. *Festuca ovina*	Buffalo grass

Answers

II. Fill in the blanks

1. 250-300 g
2. Sodding
3. 5-6 cm
4. Stolonizing
5. Dethatching

III. Choose the best answer

1. Tall fescue
2. *Dichondra repens*
3. Bermuda grass
4. Sheep fescue
5. Fine fescue
6. Creeping bent grass
7. Kentucky blue grass
8. Fine fescue
9. Zoysia grass
10. Centipede grass
11. Bermuda grass
12. Buffalo grass
13. Tall fescue
14. Kentucky blue grass
15. Creeping bent grass
16. Velvet bent grass
17. Tall fescue
18. Fine fescue
19. Tall fescue
20. Dollar spot

III. Match the following

1. Creeping bent grass
2. Tall fescue grass
3. Rye grass
4. Dandelion
5. Bent grass
6. Perennial rye grass
7. Buffalo grass
8. *Agrostis* spp
9. *Stenotaphrum secundatum*
10. Sheep fescue

ANNEXURE VIII

TURF GRASSES

S. No.	Botanical Name	Common name	Leaf texture	Establishment rate	Remarks
1.	*Agrostis alba*	Red top	Medium	Fast	Less tolerant to shade, moderately tolerant to drought, cold, salinity and heat, requires low-medium fertilizers. With moderate wear tolerance
2.	*Agrostis canina*	Velvet bent grass	Fine	Fast	Highly tolerant to shade, cold and requires medium fertilizers. With moderate wear tolerance
3.	*Agrostis palustris*	Creeping bent grass	Fine	Medium-Fast	Moderately tolerant to shade, moderately tolerant to drought, heat, highly tolerant to cold and salinity, requires high fertilizers. With poor wear tolerance
4.	*Agrostis tenuis*	Colonial bent fine grass	Fine	Medium-fast	Moderately tolerant to shade, cold, drought, heat and very sensitive to salinity, requires high fertilizers. With poor wear tolerance

5.	*Axonopus affinis*	Carpet grass	Coarse	Slow-medium	Poor tolerance to shade, drought, salinity and moderately tolerant to cold, requires low fertilizers with poor wear tolerance.
6.	*Axonopus compressus*	Broadleaf carpet grass	Coarse	Slow-medium	It is common on roadsides and in gardens, waste areas and plantations. It is best adapted to moist, warm environments and is moderately shade tolerant, although it grows well in full sunlight. It cannot withstand waterlogged conditions for long periods but grows on a wide range of soils, from sandy to heavy clay loams.
7.	*Cynodon dactylon*	Bermuda grass	Fine-medium	Very fast	Drought and cold tolerant, requires high fertilizers. With good wear tolerance. Hariyali and Calcuttia are common selections.
8.	*Dichondra repens*	Dichondra	-	-	It does not tolerate freezing and is easily killed by cold temperatures. It shows good resistance to diseases.

9.	*Eremochloa ophiuroides*	Centipede grass	Medium-coarse	Slow-medium	Very sensitive to salinity, cold, drought and moderately tolerant to shade, requires very low fertilizers, with very poor wear tolerance.
10.	*Festuca arundinacea*	Tall fescue	Medium-coarse	Slow	Good for high hilly areas in orchards, requires medium fertilizers.
11.	*Festuca elatior*	Meadow fescue	Medium	Slow	Not common lawn grass.
12.	*Festuca ovina*	Sheep fescue	Fine	Slow-medium	Good for high hilly areas, requires medium fertilizers.
13.	*Festuca rubra*	Red creeping fescue	Fine	Medium	Highly tolerant to shade and moderately tolerant to drought, cold, heat and wear, requires low fertilizers.
14.	*Lolium multiflorum*	Annual rye grass	Medium	Medium	Not very common lawn grass
15.	*Lolium perene*	Perennial rye grass	Medium	Fast	Low tolerance to drought, shade, cold and moderate tolerance to salinity and high wear tolerance.
16.	*Ophiopogan japonicus*	Mondo grass	-	-	Creates a lush groundcover of dense, dark green grass-like clumps. This dwarf is only half the size of the species. Excellent as an edging plant or tucked into

					rocks for a pleasing contrast.
17.	*Paspalum notatum*	Bahia grass	Coarse	Medium-slow	Moderately tolerant to drought, shade, cold, heat, salinity and wear, require low fertilizers.
18.	*Paspalum vaginatum*	Seashore paspalum	Coarse – very fine	-	It is propagated by sod or stolons. It is known for its high salinity tolerance and some of the varieties can be maintained with seawater containing 34, 000 pp salt.
19.	*Poa annua*	Annual blue grass	Fine-medium	Slow	Not commcn lawn grass.
20.	*Poa compressa*	Canada blue grass	Medium	Slow-medium	Not common lawn grass, poor wear tolerance.
21.	*Poa pratensis*	Kentucky blue grass	Fine-medium	Medium	Low tolerance to shade, salinity and moderate tolerance to drought, heat and salinity, requires slow-medium fertilizers.
22.	*Poa trivialis*	Rough blue grass	Fine-medium	Slow	Highly tolerant to cold and moderately tolerant to shade, drought, heat and salinity, requires lowmedium fertilizers.
23.	*Stenotaphrum secundatum*	St. Angustine	Very Coarse	Fast	Highly tolerant to shade and moderately tolerant to

					drought, salinity and wear, but low very tolerance to cold and requires high fertilizers.
24.	*Zoysia japonica*	Japanese/Korean grass	Fine	Slow	Moderately tolerant to drought, cold, shade and heat, require medium fertilizers. With good wear tolerance.
25.	*Zoysia matrella*	Manilla grass	Fine	Fast	Moderately tolerant to drought, cold, shade and heat, require medium fertilizers. With good wear tolerance.
26.	*Zoysia tenuifolia*	Korean velvet grass/ Mexican grass	Fine	Medium	Moderately tolerant to drought, cold, shade and require medium fertilizers. With very good wear tolerance.

Chapter **14**

Ferns

1. **Ornamental ferns**

Adiantum capillus – Maiden hair fern

Asplenium nidus – Bird's nest fern

Cheilanthes argentea – Silver fern

Davallia bullata – Squirrels foot fern

Nephrolepsis aureum - Boston fern

Polypodium aureum – Hare's foot fern

Pteris cretica – Brake/ Ribbon fern

Drynaria quercifolia -Oak-leaf fern / basket fern

2. **Ornamental grasses**

Andropogon gerardii – Big bluestem

Chasmanthium latifolium- Indian wood-oats

Cortaderia selloana – Pampas grass

Liriope muscari – lily turf

Muhlenbergia capillaries – Pink muhly grass, hair grass

Muhlenbergia lindheimeri – Blue muhly grass

Ophiopogon japonicus –Mondo grass

Panicum amarum – Beach panic grass

Pennisetum orientale – Oriental fountain grass

Pennisetum setaceum var. rubra – Red fountain grass

Spartina bakeri – Sand cord grass

Tripsacum dactyloides – Fakahatchee grass

ANNEXURE IX

FERNS

S. No.	Botanical Name	Common name	Family
1.	*Adiantum capillus – veneris*	Venus - hair	Polypodiaceae
2.	*Adiantum decorum*	-	Polypodiaceae
3.	*Adiantum macrophyllum*	-	Polypodiaceae
4.	*Adiantum pedatum*	-	Polypodiaceae
5.	*Adiantum raddianum*	Delta maiden-hair	Polypodiaceae
6.	*Adiantum tenerum*	Fan maiden-hair	Polypodiaceae
7.	*Adiantum trapeziforme*	-	Polypodiaceae
8.	*Aslenium bulbiferum*	Mother fern	Polypodiaceae
9.	*Aslenium nidus*	Bird'd nest fern	Polypodiaceae
10.	*Blechnum gibbbum*	-	Polypodiaceae
11.	*Cyrtomium caryotiedeum*	-	Polypodiaceae
12.	*Cyromium falcatum*	Holly-fern/Fish tail fern	Polypodiaceae
13.	*Davallia bullata*	-	Polypodiaceae
14.	*Davallia canariensis*	Hare's-foot fern	Polypodiaceae
15.	*Davallia fijiensis*	Rabbit's-foot fern	Polypodiaceae
16.	*Dicksonia fibrosa*	Golden tree fern	Dicksoniaceae
17.	*Doryopteris ludens*	-	Polypodiaceae
18.	*Nephrolepis biserrata furcans*	Fish-tail fern	Polypodiaceae
19.	*Nephrolepis exaltata*	Sword fern	Polypodiaceae
20.	*Pityrogramma calomelanos*	Silver fern	Polypodiaceae
21.	*Pityrogramma chrysophylla*	Golden fern	Polypodiaceae
22.	*Platycerium alcicorne*	Elkhorn fern	Polypodiaceae
23.	*Platycerium diversifolium*	-	Polypodiaceae
24.	*Platycerium willinckii*	Silver staghorn	Polypodiaceae

25. *Polypodium aureum*	Hare's foot fern	Polypodiaceae
26. *Polypodium percussum*	-	Polypodiaceae
27. *Polypodium phyllitidis*	Strap fern	Polypodiaceae
28. *Polypodium punctatum*	-	Polypodiaceae
29. *Polystichum acrostichoides*	Christmas fern/shield fern	Polypodiaceae
30. *Polystichum aristatum*	East Indian holly-fern	Polypodiaceae
31. *Polystichum lobatum*	-	Polypodiaceae
32. *Polystichum setiferum*	Hedge fern	Polypodiaceae
33. *Polystichum setosum*	Bristle fern	Polypodiaceae
34. *Pteris cretica*	-	Polypodiaceae
35. *Pteris ensiformis*	Victoria fern	Polypodiaceae
36. *Pterismultifida*	-	Polypodiaceae
37. *Pteris quadriaurita*	-	Polypodiaceae

Chapter **15**

Rock Garden

I. General information

1. **Cacti plants** suitable for rock garden are Opuntia, Cereus and Mamillaria
2. **Succulents** suitable for rock garden are Agave, Aloe, *Euphorbia splendens, Sedum, Sempervivum, Stapelia, Yucca, Furcraea*
3. **Ferns** suitable for rock garden are *Drynaria, Nephrodium, Nephrolepis, Polypodium*
4. **Famous rock** garden of India is located at Chandigarh
5. The founder of Rock garden in Chandigarh is Nek Chand
6. **Annuals** suitable for rock garden are
   - *Alyssum*
   - *Antirrhinum*
   - *Bellis perennis*
   - *Brachycome iberidifolia*
   - *Browallia elata*
   - *Salvia splendens*
   - *Viola tricolor hortensis*
   - *Candytuft\Dianthus chinensis*
   - *Hymenatherum tenuifolium*
   - *Mesembryanthemum criniflorum*
   - *Phlox*
   - *Verbena hybrida*
   - *Zinnia linearis*
7. **Shrubs** suitable for rock garden are
   - *Adeneum obesum*
   - *Asclepias curassavica*
   - *Asystasis coromandeliana*
   - *Azalea*
   - *Brya ebenus*
   - *Calliandra inaequilatera*
   - *Callistemon lanceolatus*
   - *Cassia alata*

- *Cassia biflora*
- *Cassia laevigata*
- *Catesbaea spinosa*
- *Cotoneaster* sp.
- *Crossandra* sp.
- *Clerodendron macrosiphon*
- *Cuphea* sp.
- *Cystisus* sp.
- *Daphne* sp.
- *Duranta plumieri variegate*
- *Euonymus sp.*
- *Jatropha podagrica*
- *Juniperus chinensis*
- *Juniperus horizontalis*
- *Lantana sellowiana*
- *Muehlenbeckia platyclados*
- *Ruellia* sp.
- *Russelia juncea*
- *Sanchezia nobilis variegata*
- *Thuja* sp.
- *Vaccinium* sp.

8. **Herbaceous and bulbous** plants suitable for rock garden are

- *Angelonia grandiflora*
- *Aster amellus*
- *Aster novae-angliae*
- *Crocus*
- *Dracaena sanderiana*
- *Fittonia argyroneura*
- *Impatiens holstii*
- *Impatiens sultanii*
- *'Oxalis rubra*
- *Portulaca*
- *Sansevieria cylindrica*
- *Sansevieria zeylanica*
- *Setcreasea*
- *Tradescantia zebrine*
- *Verbena erinoides*
- *Vinca rosea*
- *Vittadinia australis*
- *Zephyranthus* sp.

II. Choose the best answer

1. Cacti plant suitable for rock garden is......................

   Mamillaria *Echinocactus* *Echinopsis*

2. Famous rock garden of India is located at......................

   Tamil Nadu Madhya Pradesh Chandigarh

3. Succulent suitable for rock garden is......................

   Crassula *Yucca* *Adenium*

4. The founder of Rock garden at Chandigarh is.......................

   Hyder Ali Fadai Khan Nek Chand

5. Fern suitable for rock garden is.......................

   Davallia *Nephrolepis* *Cheilanthes*

Answers

1. Mamillaria
2. Chandigarh
3. *Yucca*
4. Nek Chand
5. *Nephrolepis*

Chapter **16**

Water Garden

I. General information

1. Lotus stem is referred as naval cord of Lord Vishnu
2. Lotus flower is described as the throne of Brahma
3. Vishnu, Brahma and Laxshmi are associated with lotus
4. Poet Asvaghosa reported that Lord Buddha saw lotuses and flowering trees in Nandavana
5. **Surface flowering plants of water garden** are

- *Aponogeton distachys*
- *Eurale ferox*
- *Limnanthemum indica*
- *Monochoria hastifolia*
- *Nelumbo lutea*
- *Nelumbo nucifera*
- *Nymphaea tetragona*
- *Orontium aquaticum*
- *Victoria regia*

6. **Oxygenator plants of water garden** are

- *Callitriche autumnalis*
- *Ceratophylum spp.*
- *Elodea Canadensis*
- *Ranunculus aquatilis*
- *Vallisneria spiralis*

7. **Floating plants of water garden** are

- *Azolla caroliniana*
- *Eichornia speciosa*
- *Lemna gibba*
- *Trapa narans*

8. **Marginal plants of water garden** are

- *Acorus calamus*
- *Calla palustris*
- *Cyperus alternifolius*
- *Iris pseudocorus*

- *Iris laevigata*
- *Saagittaria sagittifolia*
- *Typha angustifolia*

9. *Aponogeton distachys* is commonly called as Cape pond weed
10. *Limnanthemum indica* is commonly called as Water snow flake
11. *Nelumbo lutea* is commonly called as American lotus
12. *Nelumbo nucifera* is commonly called as Indian lotus
13. *Orontium aquaticum* is commonly called as Garden club
14. *Ranunculus aquatilis* is commonly called as water butter cup
15. *Vallisneria spiralis* is commonly called as tape grass
16. *Eichornia speciosa* is commonly called as water hyacinth
17. *Lemna gibba* is commonly called as Duck weeds
18. *Trapa narans* is commonly called as Water chestnut
19. *Acorus calamus* is commonly called as Sweet flag
20. *Calla palustris* is commonly called as Bog arum
21. *Iris pseudocorus* is commonly called as Yellow flag
22. *Saagittaria sagittifolia* is commonly called as Common arrowhead

II. Choose the best answer

1. An example for surface flowering plant of water garden is.....................

   Callitriche autumnalis *Ceratophylum spp.* *Victoria regia*

2. An example for floating plant of water garden.....................

   Lemna gibba *Acorus calamus* *Calla palustris*

3. Lotus flower is described as the throne of.....................

   Brahma Vishnu Shiva

4. Common name of water plant '*Calla palustris*' iscalled as.....................

   Water chestnut Bog arum water butter cup

5. Water garden plant *Acorus calamus* is commonly called as.....................

   Sweet flag Yellow flag Garden club

III. Fill in the blanks

1. *Nelumbo lutea* is commonly called as.....................
2. *Nelumbo nucifera*is commonly called as.....................
3. *Vallisneria spiralis* is commonly called as tape grass.....................
4. *Saagittaria sagittifolia* is commonly called as.....................
5. *Calla palustris* is an example of..................... type of plant used in water garden

Answers

II. Choose the best answer

1. *Victoria regia*
2. *Lemna gibba*
3. Brahma
4. Bog arum
5. Sweet flag

III. Fill in the blanks

1. American lotus
2. Indian lotus
3. Tape grass
4. Common arrowhead
5. Marginal type

Chapter 17

Home Garden

I. General information

Plants suitable for balcony gardening

- *Acalypha*spp.
- *Anthurium* spp.
- *Cryptanthus* spp.
- *Codiaeum variegatum*
- *Cordyline* spp.
- *Gynura* spp.
- *Maranta* spp.
- *Philodendron* spp.
- *Setcreasea pallida*
- *Ananas variegata*
- *Aspidistra variegata*
- *Caladium* spp.
- *Coleus* spp.
- *Dracaena* spp.
- *Manihot* spp.
- *Pandanus veitchi*
- *Polyscias* spp.
- *Xanthosoma lindenii*

Plants suitable for rock garden

- *Phlox* spp.
- *Alyssum* spp.
- Sweet William
- *Peperomia*spp.
- *Tradescantia* spp.
- Strawberry
- *Verbena* spp.
- *Dianthus*spp.
- *Chlorophytum*spp.
- Dwarf sansevieria
- *Zebrina* spp.
- Gooseberry

Plants suitable for making topiary

- *Murraya paniculata* (Kamini)
- *Ixora chinensis* (Rangan)
- *Lawsonia alba* (Mehandi)
- *Mimusops elengi* (Moulsari)

- *Duranta plumeri*
- *Tabernaemontana coronaria* (Chandni)
- *Casuarina equisetifolia*
- *Juniperus chinensis*
- *Malphigia coccigera*
- *Ficus* 'Panda'

Plants suitable for making bonsai are

- Moraceae : *Ficus bengalensis, F. elastica, F. religiosa, F. retusa, F. rumphi, F. indica, F. lyrata*
- Rutaceae : *Citrus aurantium, C. jambheri, Citrus reticulata, C. acida, C. maxima,Fortunella japonica, Limonia trifolia, Feronia limonia, Murraya exotica*
- Leguminosae : *Albizia lebbek, Delonix regia, Cassia nodosa, Tamarindus indica,Bauhinia variegate*
- Sapotaceae : *Mimusops elengi, Manilhara hexandra, Achras sapota* · Oxalidaceae : *Averrhoea carambola*
- Myrtaceae : *Psidium cattleyanum*

Palm species suitable for home gardening are

Feather leaved palms :

- *Caryota mitis*
- *Chrysalidocarpus lutescens*
- *Kentia belmoreana*
- *Oreodoxa regia*
- *Phoenix roebelenii*

Fan leaved palms :

- *Livistonia chinensis*
- *Pritchardia pacifica*
- *Raphis humilis*
- *Stevensonia grandiflora*
- *Thrinax argentea*
- *Washingtonia filifera*

Conifers suitable for home gardening are

- Araucariaceae : *Araucaria cookii, Araucaria cunninghamii*
- Pinaceae : *Pinus wallichiana* (Bhutan pine, Himayalan pine), *Pinus sylvestris* (Gold pine)
- Taxodiaceae : *Cryptomeria japonica* (Japanese ceder) · Pinaceae : *Cedrus deodara*
- Cupressaceae : *Cupressus macrocarpa* (Gold crest), *Juniperus chinensis* (Chinese juniper), *J. communic* (Common juniper), *Thuja orientalis, T. compacata*

- Graminae : *Bambusa glaucesceus, B. ventricosa, Arundinaria anceps* (Anceps bamboo), *A. murieliae* (Yellow bamboo), *Phyllostachys flexuosa* (Zigzag bamboo)
- Strelitiziaceae : *Ravenea madagascariensis* (Traveller's tree)

Plants suitable for dish (or) trough garden

- Araucaria
- Agaloenema
- *Begonia rex*
- *Calathea*
- *Coleus*
- *Peperomia*
- *Zebrina pendula*
- *Chlorophytum*
- *Hedera helix*
- *Maranta luconneura*
- *Pilea muscosa*
- *Sansevieria*
- *Scindapsus*
- *Selaginella*

Plants suitable for dry & hot conditions

- Chlorophytum
- *Cacti*
- *Aralia*
- *Ficus elastic*
- *Pedilanthus*
- *Peperomia*

Plants suitable for cool room

- *Cissus antactica*
- *Hedera helix*
- *Seteresia purpurea*

Plants for small pots

- *Zebrina pendula*
- *Pilea muscosa*
- *Impatinens sultani*
- *Cacti and Succulents*
- *Maranta luconeura*
- *Pilea cadierei*
- *Chlorophytum*
- *Nasturtium*

Hardy plants for drought condition

- *Aspidistra* sp.
- *Dieffenbachia* sp.
- *Pedilanthus* sp.
- *Chlorophytum* sp.
- *Sansevieria* sp.
- *Pandanus* sp.

- *Philodendron* sp.
- *Dracena* sp.
- *Monstera* sp.
- *Ficus elastia*
- *Scindapsus* sp.

Plants for bowl & terrarium

- *Asparagus plumosus*
- *Aglaonema*
- *Maranta luconeura*
- *Calathea illustris*
- *Begonia rex*
- *Ficus pumila*
- *Fittonia verchaffeltii*
- *Peperomia obtusifolia*
- *Philodendron*
- *Pilea muscosa*
- *Asplenium*
- *Tradescantia*
- *Selaginella*
- *Saintpaulia*
- *Zebrina pendula*

Plants for hanging basket

- *Zebrina pendula*
- *Nasturtium*
- *Chlorophytum*
- *Nephrolepsis exaltata*
- *Scindapsus aureus*
- *Begonia* sp.
- *Scindapsus*
- *Pilea muscosa*
- *Asparagus*
- *Pilea cadierei*
- *Zebrina pendula*

Climbing plants for indoor garden

- *Hedera helix*
- *Cissus antarctica*
- *Ficus pumila*
- *Scindapsus*
- *Philodendron scandens*
- *Hoya carnosa*

Plants for dark corner

- Aspidistra
- Sansevieria
- Selaginella
- Monstera
- Maranta
- Tradescantia
- Scindapsus
- *Zebrina pendula*
- *Philodendron*
- *Araucaria*

Plants for east or west window

- Araucaria
- Anthurium
- Begonia
- Caladium
- Tradescantia
- Ferns
- *Pandanus*
- *Zebrina pendula*
- *Dieffenbachia*
- *Cissus antarctica*
- *Dracena*
- *Ficus elastica*
- *Gloxinia*

Plants for South window

- Acalypha sp.
- *Cacti* sp.
- *Coleus* sp.
- *Poinsettia* sp.
- *Euphorbia milii*
- *Beloperone guttata*
- *Bromeliads* sp.
- *Lilium longiflorum*
- *Geranium*

Plants for North window

- *Aspidistra* sp.
- *Dieffenbachia* sp.
- *Chlorophytum* sp.
- *Monstera* sp.
- *Philodendron* sp.
- *Sansevieria* sp.
- *Aglaonema* sp.
- *Hedera helix*
- *Scindapsus* sp.
- *Peperomia* sp.
- *Selaginella* sp.
- *Zebrina pendula*

Plants for rock garden – Sunny condition

- Cacti
- *Daedalacanthus* sp.
- Miniature roses
- *Verbena erinoides*
- Succulents
- *Lantana* sp.
- *Setcreasea purpurea*

Rock garden – Shade and semi-shade

- Fittonia
- Selaginella
- Peperomia
- Zebrina

- Monstera
- *Rivinia*
- *Scindapsus*
- *Tradescantia*
- *Rhoeo discolor*
- *Sansevieria*
- *Ruscus*

Plants for water garden

Oxygenating plants : Sagittaria, *Vallisneria, Carlomba carolinana, Anacharis, Elodea*

Water plants: Water lilies (Nymphaea), *Nymphaea capensis* (Deep blue), *N. caerulea* (Pale blue), *N.. lotus* (Purple white or white with pink or red at base of petals, *N. pubescens* (White), *N. rubra* (Large double, deep red), *N. stellata* (medium to large flowers, pale blue), *Nelumbium speciosum, Typha latifolia, Typha angustata*

Plants for marshy areas

- Arum lily - *Calla palustria*
- *Cyperus – Cyperus alternifolius*
- *Irises – Iris foliosa*
- *Saxifraga pellata*

Moisture loving plants

- Alocasia
- Anemone
- Colocasia
- Hedychium
- Hemerocallis
- Kniphofia (red – hot pocker)
- Mimulus
- Ranunculus
- Saxigraga

- Azalea Daphne
- Hydrangea
- Maple (Acer)
- Rhododendron
- *Sambucus niger*
- Spiraca

Ornamental grasses & bamboos also look pretty by water side

- Pampas grass – *Cortaderia argentea*
- *Pennisetum longistylum*
- *Phyllostachys*
- *Broom grass – Thysanolaena maxima*

Propagation of House plants

- **Cuttings** – *Rex begonia, Gloxinia, Bryophyllum, Peperomia, African violet*
- **Root division** – Maranta, Dieffenbachia, Bellis, Gerbera, Kniphofia, Chrysanthemum
- Ground layering – Jasmine, Hydrangea, Azalea. Climbing rose, *Vinca major*, Clematis, Hoya
- **Air layering** – Bougainvillea, Magnolia, Rhododendron, Croton, Coleus, Excoecaria, Ficus, Camellia, Hoya
- **Budding** - Rose
- **Stem cuttings** – Dieffenbachia, Acalypha, Aralia, Begonia, Dracaena, Coleus, Cordyline, Monstera, Polyscias, Crasula, Epipremnum, Excoecaria, Ficus, Fittonia, Gynura, Hedera, Irescine, Leea, Manihot, Oplismenus, Peperomia, Philodendron, Pleomele, Pseuderanthemum, Sanchezia, Schefflera, Scindapsus, Setcreasea, Tradescantia, Cebrina, Beloperone, Euphorbia, Gloxinia, Graptophyllum, Hoya, Hypoestes, Impatiens, Pelargonium, Aphelandra, Clerodendrum, Columnea, Episcia, Vanilla, Pinus, Juniperus, Thuja, Bambusa
- **Soft wood cuttings** – Coleus, Impatiens, Peperomia, Schefflera, Strobilanthes
- **Root cuttings** - Pleomele
- **Vine cutting** – Syngonium, Scindapsus, Philodendron
- **Leaf cuttings** – Peperomia, Scindapsus, Sansevieria, Croton, Gloxinia
- **Leaf blade cutting** - Bryophyllum
- **Runners** – Tradescantia, Zebrina, Philodendron, Epipremnum, Saxifraga
- **Division** – Asparagus, Aspidistra, Begonia, Calathea, Chlorophytum, Sansevieria, Cryptanthus, Cyperus, Fittonia, Helxine, Kaempferia, Maranta, Ophiopogon, Oplismenus, Pilea, Clematis, Gloxinia,

Spathiphyllum, Hedychium, Heliconia, Oxalis, Bambusa, Orchids

- **Seed** – Asparagus, Begonia, Chlorophytum, Coleus, Ficus, Nolina, Pilea, Rhoeo, Gloxinia, Hypoestes, Impatiens, Senecio, Gloriosa, Hedychium, Palms, Cycas, Araucaria, Pinus, *Juniperus,* Thuja, Bambusa, Ravenala
- **Offsets** Eucharis, Haemanthus, Hemarocallis, Strelitzia, Chlorophytum, Dieffenbachia, Nolina, Pandanus, Rhoeo, Biigergia, Neoregelia
- **Bulbils** – Furcraea, Cycas
- **Tissue culture** - Anthurium
- **Spores** - Aspidium
- **Suckers / Slips / by rooting leafy crown** - Ananas
- **Division of tubers** - Caladium
- **Tubers** - Gloriosa

Ferns suitable for home gardening

- *Dryopleris filix-mas* (Male fern)
- *Dicksonia antarctica* (Australian tree fern)
- *Polystichum aculeatum*
- *Microlepia strigosa*
- *Selaginella strigosa*
- *Polystichum setiferum*
- *Matteuccia struthiopteris*
- *Polypodium vulgare*
- *Adiantum venustum*

Cacti suitable for home gardening

- *Astrophytum myriostigma*
- *Cephalocereus senilis* – Oldman Cactus
- *Echinocactus grusonii* - Golden Barrel Cactus
- *Ferocactus latispinus*

- Lophophora williamsii
- *Mammillaria bocasana*
- *Opuntia microdasys*
- *Chamaecereus silvestrii*
- *Gymnocalycium mihanovichii*
- *Melocactus concinnus*
- *Notocactus graessneri*
- *Notocactus leminghausii*

Succulents suitable for home gardening

Agave victoria – reginae	Agavaceae
Agave filifera	
Agave stricta	
Agave americana	
Haworthia chalwinii	Liliaceae
Haworthia reinwardtii	
Aloe humilis	
Aloe aristata	
Aloe ferox	
Aptenia cordifolia	Aizoaceae
Aridaria splendens	
Bergeranthus scapigera	
Caralluma burchardii	Asclepiadaceae
Ceropegia juncea	
Duvalia elegans	
Echinopsis dammaniana	
Kleinia acaulis	Compositae
Kleinia neriifolia	
K. tomentosa	
Dyckia sulphurea	Bromeliaceae
Dyckia rariflora	
Crassula arborescens	Crassulaceae
Crassula falcata	
Crassula perfoliata	

❐ *Kalanchoe beharensis*	
❐ *Kalanchoe tomentosa*	
❐ *Kalanchoe marmorata*	
❐ *Cotyledon undulata*	
❐ *Greenovia aurea*	
❐ *Euphorbia bupleurifolia*	Euphorbiaceae
❐ *Euphorbia grandidens*	
❐ *Euphorbia horrida*	
❐ *Euphorbia polygona*	
❐ *Euphorbia obesa*	
❐ *Euphorbia ingens*	
❐ *Euphorbia valida*	
❐ *Anacampseros papyracea*	Portulaceae
❐ *Anacampseros ustulata*	
❐ *Anacampseros alstonii*	
❐ *Anacampseros rufescens*	
❐ *Portulacaria afra*	

Flowering trees suitable for home gardening

Common name	Botanical Name	Family	Remarks
Champaka	*Michelia champaca*	Magnoliaceae	Flowers orange, yellow, strongly scented, March - Sep
Indian Medlar	*Mimusops elengi*	Sapotaceae	Dull white (or) greenish white, highly fragrant, summer months
Tree of heaven	*Amherstia nobilis*	Leguminosae	Flowers brilliant red & yellow, Jan - April
-	*Jacaranda mimosaefolia*	Bignoniaceae	Flowers bluish purple, March - April
Wild chincona	*Anthocephalus cadamba*	Rubiaceae	Flowers scented, corolla orange, calyx pale greenish
Cannon ball tree	*Couroupita guianensis*	Lecythidaceae	Flowers scented, white, yellow & pink (or) crimson
Iron wood	*Mesua ferrea*	Clusiaceae	Flowers white, fragrant
Laurel Magnolia	*Magnolia grandiflora*	Magnoliaceae	Flowers white, March– May
Bayur tree	*Pterospermum acerifolium*	Sterculiaceae	Slow growing, Flowers scented & pure white petals
Asoka tree	*Saraca asoca*	Leguminosae	Yellow and orange colour flowers
Buddhist Bauhinia	*Bauhinia variegata*	Caesalpiniaceae	White flowers, Late Jan – Mid March
Pagoda tree	*Plumeria acuminata*	Apocynaceae	White with yellow centre, scented
	Plumeria acutifolia		-
	Plumeria rubra		-

Peacock flowers	*Delonix regia*	Leguminosae	Flowers deep crimson, scarlet (or)orange
Coral jasmine	*Nyctanthes arbortristis*	Oleaceae	Scented, white flowers with orange corolla tube
Trumpet tree	*Tabebuia pallida*	Bignoniaceae	Cuban pink trumpet tree, flowers light pink
	Tabebuia rosea		Flowers rosy pink, Feb - March

Indoor plants for flowers

Common name	Family	Botanical name	Special features
Anthurium	Araceae	*Anthurium andreanum var.rhodochlorum*	Spathe salmon red with green extremities
		*Anthurium andreanum var.*giganteum	Large salmon red spathes
		*Anthurium andreanum var.*alba	Spathes white
		Anthurium andreanum var. roseum	Spathes shining
Aphelandra	Acanthaceae	*Aphelandra aurantiaca*	Flower head orange
		Aphelandra pectinata	Scarlet
		Aphelandra squarrosa	Flower head yellow with greenish – yellow bracts
		Aphelandra tetragona	Scarlet
Clerodendrum	Verbenaceae	*Clerodendrum thompsoniae*	Bright crimson flower with clear white

		(Syn. C. balfouri)	bracts
Columnea	Gesneriaceae	*Columnea banksii*	Scarlet
		Columnea gloriosa	Scarlet & yellow
		Columnea gloriosa var. purpurescens	Scarlet & Yellow
		Columnea microphylla	Scarlet & yellow
Episcia	Gesneriaceae	*Episcia* sp.	Scarlet
Haemanthus	Amaryllidaceae	*Haemanthus katherinae*	Scarlet on separate stalks forming a dense head
		Haemanthus magnificus	Large head of orange scarlet flowers with golden stamens
Hedychium	Zingiberaceae	*Hedychium coronarium*	Plant resembling canna, flowers are white, strongly scented
Heliconia	Heliconiaceae	*Heliconia metallica*	Flowers red enclosed in narrow boat-shaped green bracts
		Heliconia striata	Flowers are long and orange
		Heliconia bihai	Flowers greenish yellow, bracts boat shaped, burnt red with yellow tip, the edges apple green to yellow
Day Liliy	Liliaceae	*Hemarocallis aurantiaca*	Japanese day lilies – Flowers orange yellow
		Hemarocallis citrina	Flowers lemon yellow, slightly
		Hemarocallis citrine var. citrina	Flowers are citron yellow

		Hemarocallis flava	Flowers orange yellow
Hoya	Asclepiadaceae	*Hoya bella*	Flowers waxy white with a purple centre
		Hoya carnosa	Flowers pinkish white, fragrant
Oxalis	Oxalidaceae	*Oxalis hedysaroides*	Flowers bright yellow
BOP (Bird - of -Paradise)	Strelitziaceae	*Strelitzia reginae*	Sepals brilliant orange yellow, spathe boat shaped
Vanilla	Orchidaceae	*Vanila arometica*	Yellowish in bunches, fragrant

Indoors plants for foliage

Common name	**Family**	**Botanical name**	**Special features**
Acorus	Araceae	*Acorus calamus*	-
Aglaonema	Araceae	*Aglaonema angustifolia*	Very short stature
		Aglaonema crispum	Large pointed leathery leaves, slightly grayish green
		Aglaonema modestum	Chinese evergreen thrives in poorly lighted place, found to grow in bottles filled with only water
		Aglaonema pseudobracteatum	Deep green leaves variegated light green & yellow centre. very attractive & hardy
		Aglaonema 'Silver Queen'	Leathery lanceolate leaves with greenish silvery blotches on deep green background. Very bushy with a large clump

Alocasia	Araceae	*Alocasia argyraea*	Leaves dark green with silver sheen
		Alocasia cuprea	Leaves wide, purple beneath, dark metallic green above
		Alocasia indica	Very tall, green
		Alocasia indica var. Metallica	Deep purple leaves with metallic sheen
		Alocasia indica var. Varigata	Leaves and stalks mottled white with violet patches
		Alocasia 'Hilo Beauty'	Very attractive, thick leaves nicely marked with irregular areas of coloured, translucent blotches
		Alocasia 'Korthalsii'	Leaves long, deep olive green with white mid rib and veins, purple underneath
Alpinia	Zingiberaceae	*Alpinia allughas*	Flowers red
		Alpinia rafflesiana	Flowers small with red tipped golden
		Alpinia speciosa	Indian shell flower (with spots & stripes of white, purple & yellow)
Alternanthera	Amaranthaceae	*Alternanthera amoena*	Leaves green veined with orange red blotches
		Alternanthera var. Amabilis	Leaves orange – scarlet
		Alternanthera bettickiana	Leaves blotched with shades of yellow and red
		Alternanthera versicolor	Leaves coppery red
Ananas	Bromeliaceae	*Ananas comosum variegatus*	-

Anthurium	Araceae	*Anthurium bakeri*	-
		Anthurium crassinervum	Lamina elliptic to oblanceolate
		Anthurium crystallinum	Leaves broadly ovate, enlarged green leaves, ribs bold, silvery
		Anthurium digitatum	A climbing type
		Anthurium magnificum	Leaves white & green
		Anthurium veitchii	Leaves evergreen with corrugated margins
Aralia	Araliaceae	*Aralia* sp	-
Asparagus	Liliaceae	*Asparagus densiflorus* 'Myers'	Plume like shoot like foxtail. the needle like foliage is deep green
		Asparagus densiflorus 'Sprengeri'	Stems are semi-prostrate, which grows to a height of 30cm. Branches are spreading over a diameter of 90cm. Wiry shoots bear cluster of bright green cladode. True leaves are modified into spines
		Asparagus plumosus	Feather triangular leaves are extremely beautiful, that spread horizontally. Stems are climbing with dark green colour. At nodes aerial roots develop as in bamboo. A leaf is tied to a button-hole flower
		Asparagus racemosus	Woody climber with heavy fasciculated roots. The massed cladode look very beautiful

Aspidistra	Liliaceae	*Aspidistra elatior*	Parpour palm, cast-iron plant
Beaucarnea (Ponytail palm)	Asparagaceae	*Beaucarnea recurvata Syn – Nolina recurvata*	Evergreen perennial growing to 15 feet 6 inches (4.72 m) with a noticeable expandedcaudex, for the purpose of storing water inside
Bignonia	Bignoniaceae	*Bignonia masoniana*	Iron cross. The large roundish, firm green leaves are marked with palmately bands in the center
		Bignonia rex 'Silver Queen'	Painted – leaf begonia, leaves flat, soft silvery grey exception along the veins in centre & margin are metallic green
Calathea	Marantaceae	*Calathea insignis*	Silvery grey with bright green blotches linear, erect with wavy margin underside purple
		Calathea lindeniana	Dark green with emerald green zone, underside maroon
		Calathea ornata	Leaves dark green with pink (or) cream lines, dark purple beneath
		Calathea zebrina	Deep velvety green leaves yellow – green lateral veins
Chlorophytum	Liliaceae	*Chlorophytum comosum*	-
		Chlorophytum elatum	-
		Chlorophytum elatum var. 'Variegatum'	-

Crotons	Euphorbiaceae	*Codiaeum variegatum* var. 'Andreanum'	Yellow
		Codiaeum variegatum var. 'Chelsonii'	Cream, green & red
		Codiaeum variegatum var. 'tricolor'	Cream, green & gold
		Codiaeum variegatum var. 'pictum'	Large broad leaves with yellow, red & green variegation
		Codiaeum variegatum var. 'Bangkok'	Leaves roundish with yellow margins
		Codiaeum variegatum var. 'Excellent'	Dwarf with attractive, lobed foliage
		Codiaeum variegatum var. 'Iceton'	Small leaves having cream, pink, green & copper variegation
		Codiaeum variegatum var. 'Philip Geduldig'	Compact plant with broad lamina
		Codiaeum variegatum var. 'Trilobed'	Trilobed leaves variegated yellow & greens
		Codiaeum variegatum var. 'Indian Prince'	Very long slender leaves of screw – shaped with green, yellow & crimson variegation
Coleus	Labiatae	*Coleus blumei*	Very dwarf plant of different colour variegation
		Coleus frederici	Annual (or) biennial
Cordyline	Liliaceae	*Cordyline australis*	Stems long, slender terminated with a

'Purpurea'	rosette of linear leaves of purple colcur
Cordyline australis 'Cuprea'	Leaves are copper
Cordyline australis 'Doucetii'	Pink stripes along the margin of leaves. Green portion may have white streaks
Cordyline terminalis	Leaves long, broad. Tender leaves are bright reddish pink. Older leaves are coppery green.
Cordyline terminalis 'Angusta'	Leaves long, slender, colour coppery green. Red margins underneath purple
Cordyline terminalis 'Bicolor'	Leaves green with pink stripes along the margins
Cordyline terminalis 'Madame Eugene Andre'	Bushy with broad, coppery green leaves having red margin. Old leaves drop after the emergence of flowers

Annuals suitable for home gardening

S.No.	Common Name	Botanical Name	Family
1.	Balsam	*Impatiens balsamina*	Balsaminaceae
2.	Cosmos	*Cosmos bipinnatus*	Compositae
3.	Blanket flower	*Gaillardia pulchella*	Compositae
4.	Marigold	*Tagets erecta*	Asteraceae
5.	Petunia	*Petunia hybrida*	Solanaceae
6.	Moss rose/Table rose	*Portulaca grandiflora*	Portulaceae
7.	Verbena	*Verbena hybrida*	Verbenaceae
8.	Zinnia	*Zinnia elegans*	Compositae
9.	Geranium	*Pelargonium* spp.	Geraniaceae

Bulbou sornamentals suitable for home gardening

S.No.	Common Name	Botanical Name	Family
1.	Tuberose	*Polianthes tuberosa*	Amaryllidaceae
2.	Gerbera	*Gerbera jamesonii*	Asteraceae
3.	Amaryllis	*Amaryllis belladona*	Amaryllidaceae
4.	Foot ball lily	*Haemanthus* spp.	Amaryllidaceae
5.	Indian shot	*Canna indica*	Cannaceae
6.	Glory lily	*Gloriosa superba*	Liliaceae
7.	Lotus	*Nelumbo lutea*	Nymphaeaceae
8.	Water lily	*Nymphea* sp.	Nymphaeaceae
9.	White ginger lily	*Hedychium coronarium*	Zingiberaceae

Shrubs suitable for home gardening

S.No.	Common Name	Botanical Name	Family
1.	Allamanda	*Allamanda neriifolia*	Apocynaceae
2.	Queen of night	*Cestrum nocturnum*	Solanaceae
3.	China rose	*Hisbiscus rosa-sinensis*	Malvaceae
4.	Mussanda	*Mussanda erythrophylla*	Rubiaceae
5.	Chinese Ixora	*Ixora chienesis*	Rubiaceae
6.	Crape Myrtle	*Lagerstroemia indica*	Lythraceae
7.	Egyptian Stay cluster	*Pentas lanceolata*	Rubiaceae

8.	Coral plant	*Russelia juncea*	Scrophulariaceae
9.	Crepe jasmine	*Tabernaemontana coronaria*	Apocynaceae
10.	Jasmine	*Jasminum* spp.	Oleaceae
11.	Cape jasmine	*Gardenia jasminoides*	Rubiaceae
12.	Orange jasmine	*Murraya paniculata*	Rutaceae
13.	Oleander/Nerium	*Neriumindicum*	Apocynaceae
14.	Dwarf powderpuff	*Calliandra emarginata*	Mimosaceae

Plants used for group arrangement

- *Selaginella*
- *Euphorbia milii*
- *Calathea zebrine*
- *Syngonium podophyllum*
- *Pilea cadierei*
- *Sansevieria*

Plants which can be used as specimen plants

- Dieffenbachia
- Aralia
- Philodendron
- Aglaonema
- Sansevieria

Plants which can be used as large specimen plants

- *Ficus benjamina*
- *Aralia elegantissima*
- *Howea forsteriana*
- *Ficus elastic*
- *Araucaria excels*

Climbers and Creepers used for home garden are

- *Monstera deliciosa*
- *Syngonium podophyllum*
- *Hedera helix*
- *Antigonon leptopus*
- *Quisqualis indica*
- *Allamanda cathartica*
- *Cissus antarctica*
- *Ficus pumila*
- *Clematis gouriana*
- *Artabotrys odoratissimus*
- *Bignonia* spp.

Plants suitable for Window box gardening

Sunny place – Winter annuals - Aster, Phlox, Petunia, Dianthus, Pansy, Verbena and Nasturtium

Summer & rainy season annuals– Dwarf sunflower, Balsam, Zinnia, Zephyranthus

Semi-shady place– Dracaena, Dieffenbachia, Maranta, Aglaonema, Caladium, Coleus, Codiaeum, Ficus elastic

Shady place – Sciandapsus, Pedilanthus, Philodendran, tradescantia

Plants for bottle garden

- *Aglaonema commutatum*
- *Zebrina pendula*
- *Cordyline terminalis*
- *Asplenium nidus*
- *Dracaena sanderi*
- *Peperomia obtusifolia*
- *Fittonia verschafelti*
- *Selaginella*

Plants for dish or bowl garden

- Cacti – Rebutia, Lobivia, Notocactus, Echinopsis, Mammillardia
- Succulents – Kalanchoe, Bryophyllum, Sedum, *Aloe variegata*

Plants for door-steps

- Aralia
- Phildendron
- *Ficus elastic*
- Codiaeum
- Monstera
- *Howea forsteiana*

Plants for entrance of a home

- Aralia
- *Chamoedera elegans*
- *Codiaeum variegatum*
- *Calathea zebrine*
- *Dieffenbachia picta*
- *Monstera deliciosa*

Plants for kitchen room

- Sansevieria
- *Aspidistra lucida*

Plants for verandah

1. Reflected sunlight for 3-4 hours

- *Synconium podophyllum*
- *Polypodium aureum*
- *Dracena sanderiana*
- Alocasia
- Codiaeum
- Aglaonema
- *Hedera helix*
- *Maranta leuconeura*
- Peperomia
- Philodendron
- Aralia

2. Direct sunlight for 3-4 hours

- *Coleus blumei*
- *Setcreasea purpurea*
- *Tradescantia fluminensis*
- *Codiaeum variegatum* var. pictum
- *Cordyline terminalis*
- *Strobilanthes*

3. Dark and shade corner

- Sansivieria
- *Aspidistra lucida*

Plants for balcony (Area which receives direct sunlight for few hours)

- *Coleus blumei*
- *Ficus elastica* var. decora
- *Pilea cadierei*
- *Gloxinia*
- *Codiaeum variegatum*
- *Opuntia microdasys*
- *Impatiens sultani*
- *Kalanchoe*

Plants for side walls

- Chlorophytum
- Nasturtium
- *Pilea cadieri*
- Saintpaulia
- Petunia

Plants for dining table

- Small sized pot plants – *Fittonia verschafeltii, Asplenium nidus,* Scindapsus
- Glass container with clean water - Scindapsus

Plants for hall

❐ *Aspidistra lucida*	Dark corner
❐ *Sansevieria trifasciata var. laurentii*	
❐ *Anthurium magnificus*	
❐ *Aglaonema*	
❐ *Cissus antartica*	Shady areas
❐ *Ficus benjamina*	
❐ *Dieffenbachia picta*	
❐ *Feathered leaved palm*	
❐ *Saintpaulia*	Windowsill
❐ *Poinsettia*	
❐ *Coleus*	
❐ *Cacti*	
❐ *Bonsai*	

Plants for windows which receive indirect (or) reflected sunlight

- Syngonium
- Agalaonema
- Monstera
- Sansevieria
- Dieffenbachia
- Calathea
- Maranta
- Philodendron

Plants for windows which receives direct sunlight for few hours

- *Scindapsus aureus*
- Kalanchoe
- Impatiens

Plants for drawing room

- *Philodendron scandens*
- *Anthurium magnificus*
- *Maranta bucolara*
- *Caladium bicolor*

Plants for dark corner

- Sansevieria
- *Victoria dracaena*
- Aglaonema

Group plants for corner

- *Calathea zebrine*
- *Pilea cadierei*
- *Syngonium podophyllum*

Plants for reading room

- Juniperus
- Sansevieria as stony plant
- Asparagus

ANNEXURE- X

HOUSE PLANTS

S.No.	Botanical Name	Common name	Family
1.	*Aechmea angustifolia*	-	Bromeliaceae
2.	*Aechmea caudata*	-	Bromeliaceae
3.	*Aechmea chantinii*	American zebra plant	Bromeliaceae
4.	*Aechmea chlorophylla*	-	Bromeliaceae
5.	*Aechmea fasciata*	Silver vase	Bromeliaceae
6.	*Aechmea fulgens*	Coral berry	Bromeliaceae
7.	*Aechmea miniata*	-	Bromeliaceae
8.	*Aechmea nudicaulis*	-	Bromeliaceae
9.	*Aechmea orlandiana hybrid*	-	Bromeliaceae
10.	*Aechmea racinae*	Christmas jewels	Bromeliaceae
11.	*Aechmea recurvata*	-	Bromeliaceae
12.	*Aechmea triangularis*	-	Bromeliaceae
13.	*Aeschynanthus grandiflorus*	-	Gesneriaceae
14.	*Aeschynanthus lobbianus*	-	Gesneriaceae
15.	*Aeschynanthus marmonatus*	Zebra basket vine	Gesneriaceae
16.	*Aeschynanthus pulcher*	-	Gesneriaceae
17.	*Aeschynanthus speciosus*	-	Gesneriaceae
18.	*Aglonema commutatum*	-	Araceae

19.	*Aglonema costatum*	-	Araceae
20.	*Aglonema crispum*	-	Araceae
21.	*Aglonema marantifolium*	-	Araceae
22.	*Aglonema modestum*	-	Araceae
23.	*Aglonema nitidum*	-	Araceae
24.	*Aglonema rotundum*	Red aglonema	Araceae
25.	*Aglonema siamense*	-	Araceae
26.	*Alocasia amazonica*	-	Araceae
27.	*Alocasia cadierei*	-	Araceae
28.	*Alocasia cucullata*	Chinese taro	Araceae
29.	*Alocasia cuprea*	-	Araceae
30.	*Alocasia jhonstonii*	-	Araceae
31.	*Alocasia korthalsii*	-	Araceae
32.	*Alocasia lindenii*	-	Araceae
33.	*Alocasia lowii*	-	Araceae
34.	*Alocasia lowii grandis*	-	Araceae
35.	*Alocasia lowii hybrid*	-	Araceae
36.	*Alocasia macrorrhiza*	Giant alocasia	Araceae
37.	*Alocasia metallica*	-	Araceae
38.	*Alocasia micholitziana*	-	Araceae
39.	*Alocasia portei*	-	Araceae

40.	*Alocasia sanderae*	Kris plant	Araceae
41.	*Alocasia watsoniana*	-	Araceae
42.	*Alpinia sanderae*	Variegated ginger	Zingiberaceae
43.	*Alpinia zerumbet*	Shell ginger	Zingiberaceae
44.	*Ananas bracteatus*	Red pineapple	Bromeliaceae
45.	*Ananas comosus*	Golden rocket	Bromeliaceae
46.	*Ananas nanus*	-	Bromeliaceae
47.	*Anthurium andreanum*	Painter's palette	Araceae
48.	*Anthurium clarinervium*	-	Araceae
49.	*Anthurium crystallinum*	-	Araceae
50.	*Anthurium forgetii*	—	Araceae
51.	*Anthurium magnificum*	-	Araceae
52.	*Anthurium pedato-radiatum*	-	Araceae
53.	*Anthurium scherzerianum*	Flamingo flower	Araceae
54.	*Anthurium veitchii*	-	Araceae
55.	*Anthurium warocqueanum*	-	Araceae
56.	*Aplelandra squarrosa*	Saffron spine zebra	Acanthaceae
57.	*Ardisia crispa*	Coral berry	Myrsinaceae
58.	*Ardisia japonica*	-	Myrsinaceae
59.	*Asparagus asparagoides myrtifolius*	Baby smilax	Liliaceae
60.	*Asparagus densiflorus*	-	Liliaceae

61.	*Asparagus plumosus*	Asparagus fern	Liliaceae
62.	*Asparagus scandens*	Basket asparagus	Liliaceae
63.	*Asparagus setaceus*	Cypress asparagus	Liliaceae
64.	*Asparagus tetragonas*	-	Liliaceae
65.	*Aspidistra elatior*	Cast-iron plant	Liliaceae
66.	*Begonia auriculata*	-	Begoniaceae
67.	*Begonia boweri*	Miniature eyelash begonia	Begoniaceae
68.	*Begonia cheimantha*	-	Begoniaceae
69.	*Begonia Cleopatra*	Maple leaf begonia	Begoniaceae
70.	*Begonia coccinea*	Angelwing begonia	Begoniaceae
71.	*Begonia deliciosa*	-	Begoniaceae
72.	*Begonia fuchsiodies*	Fuchsia flowered begonia	Begoniaceae
73.	*Begonia hybrida*	-	Begoniaceae
74.	*Begonia masoniana*	Iron cross	Begoniaceae
75.	*Begonia metallica*	-	Begoniaceae
76.	*Begonia rex*	-	Begoniaceae
77.	*Begonia semperflorens*	Wan begonia	Begoniaceae
78.	*Begonia tuderhybrida*	Tuberous begonia	Begoniaceae
79.	*Begonia venosa*	-	Begoniaceae
80.	*Begonia versicolor*	-	Begoniaceae
81.	*Billbergia decora*	-	Bromeliaceae

82.	*Billbergia nutans*	Queen's rears	Bromeliaceae
83.	*Billbergia seidelii*	-	Bromeliaceae
84.	*Billbergia vittata*	-	Bromeliaceae
85.	*Billbergia zebrine*	Zebra urn	Bromeliaceae
86.	*Brassia actinophylla*	-	Araliaceae
87.	*Bromelia balansae*	Pinuela	Bromeliaceae
88.	*Bromelia serra*	-	Bromeliaceae
89.	*Caladium bicolor*	Heart of Jesus	Araceae
90.	*Caladium hortulanum*	Fancy-leaved caladium	Araceae
91.	*Caladium humboldtii*	Miniature caladium	Araceae
92.	*Caladium picturatum*	-	Araceae
93.	*Calathea aurantiaca*	-	Marantaceae
94.	*Calathea bachemiana*	-	Marantaceae
95.	*Calathea bella*	-	Marantaceae
96.	*Calathea clossonii*	-	Marantaceae
97.	*Calathea crocata*	-	Marantaceae
98.	*Calathea insignis*	-	Marantaceae
99.	*Calathea kegeliana*	-	Marantaceae
100.	*Calathea leonii*	-	Marantaceae
101.	*Calathea leopardina*	-	Marantaceae
102.	*Calathea lietzei*	-	Marantaceae

103.	*Calathea lindeniana*	-	Marantaceae
104.	*Calathea louisae*	-	Marantaceae
105.	*Calathea magnifica*	-	Marantaceae
106.	*Calathea makoyana*	Peacock plant	Marantaceae
107.	*Calathea medio-picta*	-	Marantaceae
108.	*Calathea metallica*	-	Marantaceae
109.	*Calathea micans*	-	Marantaceae
110.	*Calathaea ornata*	-	Marantaceae
111.	*Calathea pavonii*	-	Marantaceae
112.	*Calathea picturata*	-	Marantaceae
113.	*Calathea princeps*	-	Marantaceae
114.	*Calathea roseo-picta*	-	Marantaceae
115.	*Calathea rufibara*	-	Marantaceae
116.	*Calathea stromata*	-	Marantaceae
117.	*Calathea tigrina*	-	Marantaceae
118.	*Calathea undulata*	-	Marantaceae
119.	*Calathea veitchiana*	-	Marantaceae
120.	*Calathea warscewiczii*	-	Marantaceae
121.	*Calathea wiotii*	-	Marantaceae
122.	*Calathea zebrine*	Zebra plant	Marantaceae
123.	*Callisia repens*	Turtle vine	Commelinaceae

124.	*Chlorophytum bichetii*	-	Liliaceae
125.	*Chlorophytum comosum*	-	Liliaceae
126.	*Chlorophytum laxum*	-	Liliaceae
127.	*Chrysothemis pulchella*	-	Gesneriaceae
128.	*Cordyline australis*	-	Liliaceae
129.	*Cordyline banksii*	-	Liliaceae
130.	*Cordyline indivisa*	-	Liliaceae
131.	*Cordyline stricta*	-	Liliaceae
132.	*Cordyline terminalis*	Red dracaena	Liliaceae
133.	*Cordyline speciosus*	-	Liliaceae
134.	*Costus spicatus*	Spiral ginger	Zingiberaceae
135.	*Cryptanthus acualis*	-	Bromeliaceae
136.	*Cryptanthus beuckeri*	Marbled poon	Bromeliaceae
137.	*Cryptanthus bivittatus*	-	Bromeliaceae
138.	*Cryptanthus bromeliodies*	Rainbow star	Bromeliaceae
139.	*Cryptanthus praetextus*	-	Bromeliaceae
140.	*Cryptanthus zonatus*	-	Bromeliaceae
141.	*Ctenanthe kummeriana*	-	Marantaceae
142.	*Ctenanthe lubbersiana*	-	Marantaceae
143.	*Ctenanthe openheimiana*	-	Marantaceae
144.	*Ctenanthe pilosa*	-	Marantaceae

145.	*Ctenanthe setosa*	-	Marantaceae
146.	*Curculigo capitulate*	-	Amaryllidaceae
147.	*Cylanthus bipartitus*	-	Cylanthaceae
148.	*Cyperus alternifolius*	Umbrella plant	Cyperaceae
149.	*Cuperus diffuse*	-	Cyperaceae
150.	*Cypreus papyrus*	Egyptian paper plant	Cyperaceae
151.	*Dianela tasmanica*	Blue fax lily	Liliaceae
152.	*Dieffenbachia amoena*	-	Araceae
153.	*Dieffenbachia bausii*	-	Araceae
154.	*Dieffenbachia hoffmannii*	-	Araceae
155.	*Dieffenbachia leoniae*	-	Araceae
156.	*Dieffenbachia maculata*	Spotted dumb cane	Araceae
157.	*Dieffenbachia reginae*	-	Araceae
158.	*Dieffenbachia wallisii*	-	Araceae
159.	*Dizygotheca elegantissima*	-	Araliaceae
160.	*Dracaena deremensis*	-	Dracaenaceae
161.	*Dracaena fragrans*	-	Dracaenaceae
162.	*Dracaena godseffiana*	Gold-dust dracaena	Dracaenaceae
163.	*Dracaena hookeriana*	-	Dracaenaceae
164.	*Dracaena marginata*	Rainbow plant	Dracaenaceae
165.	*Dracaena sanderiana*	-	Dracaenaceae

166.	*Epipremnum elegans*	Devil's Ivy	Araceae
167.	*Epipremnum pinnatum*	-	Araceae
168.	*Episcia cupreata*	-	Gesneriaceae
169.	*Episcia lilacina*	-	Gesneriaceae
170.	*Episcia reptans*	-	Gesneriaceae
171.	*Euonymus japonica*	-	Celastraceae
172.	*Fatshedera lizei*	Miracle plant	Araliaceae
173.	*Fatshedera japonica*	-	Araliaceae
174.	*Ficus benjamina*	-	Moraceae
175.	*Ficus cyathistipula*	-	Moraceae
176.	*Ficus diversifolia*	-	Moraceae
177.	*Ficus elastica*	Indian rubber plant	Moraceae
178.	*Ficus jacquinifolia*	West Indian laurel fig	Moraceae
179.	*Ficus lyrata*	-	Moraceae
180.	*Ficus nitida*	Indian laurel	Moraceae
181.	*Ficuspumila*	-	Moraceae
182.	*Ficus triangularis*	-	Moraceae
183.	*Fittonia verschaffeltii*	-	Acanthaceae
184.	*Guzmania cardinalis*	-	Bromeliaceae
185.	*Guzmania lingulata*	-	Bromeliaceae
186.	*Hedera canariensis arborescens*	Algerian Ivy	Araliaceae

187.	*Hedera colchica*	Persian Ivy	Araliaceae
188.	*Hedera helix*	English Ivy	Araliaceae
189.	*Heliconia humilis*	Lobster claw	Musaceae
190.	*Heliconia illustris rubricaulis*	-	Musaceae
191.	*Heliconia metallica*	-	Musaceae
192.	*Heliconia psittacorum*	-	Musaceae
193.	*Heliconia rostrata*	Hanging lobster claw	Musaceae
194.	*Hemigraphis colorata*	Red Ivy	Acanthaceae
195.	*Hoffmannia ghiesbreghtii*	Tall taffeta plant	Rubiaceae
196.	*Hoffmannia refulgens*	Quilted taffeta plant	Rubiaceae
197.	*Homalomena lindenii*	-	Araceae
198.	*Homalomena picturata*	-	Araceae
199.	*Homalomena sulcata*	-	Araceae
200.	*Homalomea wallisii*	Silver shield	Araceae
201.	*Homalomea wendlandii*	-	Araceae
202.	*Hypocyrta glabra*	-	Gesneriaceae
203.	*Hypoestes sanguinolenta*	Freckle face	Acanthaceae
204.	*Impatiens linearifolia*	Balsam	Balsaminaceae
205.	*Impatiens walleriana*	-	Balsaminaceae
206.	*Kaempferia pulchera*	Pretty Resurrection lily	Zingiberaceae
207.	*Kaempferia roscoeana*	Peacock plant	Zingiberaceae

208.	*Leea coccinea*	-	Vitaceae
209.	*Leea rubra*	-	Vitaceae
210.	*Liriope exiliflora*	Lily turf	Liliaceae
211.	*Liriope muscari*	-	Liliaceae
212.	*Liriope platyphylla*	-	Liliaceae
213.	*Maranta leuconeura erythroneura*	Rabbits tracks	Marantaceae
214.	*Marantha leuconeura kerchoveana*	-	Marantaceae
215.	*Medinilla curtisii*	-	Melastomaceae
216.	*Medinilla magnifica*	-	Melastomaceae
217.	*Miconia hookeriana*	-	Melastomaceae
218.	*Miconia magnifica*	-	Melastomaceae
219.	*Monstera deliciosa*	-	Araceae
220.	*Monstera oblique*	-	Araceae
221.	*Nautilocalyx forgetii*	Black alloplectus	Gesneriaceae
222.	*Neoregelia carolinae*	-	Bromeliaceae
223.	*Neoregelia spectabilis*	Finger-nail plant	Bromeliaceae
224.	*Nidularium fulgens*	-	Bromeliaceae
225.	*Nidularium innocentii*	-	Bromeliaceae
226.	*Nidularium regelioides*	-	Bromeliaceae
227.	*Ophiopogon jaburan*	White lily-turf	Liliaceae
228.	*Ophiopogon japonicas*	-	Liliaceae

229.	*Ophiopogon planiscapus nigrescens*	Black dragon	Liliaceae
230.	*Pandanus baptistii*	Blue screw-pine	Pandanaceae
231.	*Pandanus sanderii*	-	Pandanaceae
232.	*Pandanus veitchii*	Variegated screwpine	Pandanaceae
233.	*Pelargonium domesticum*	Real geranium	Geraniaceae
234.	*Pellionia daveauana*	-	Urticaceae
235.	*Pellionia pulchra*	-	Urticaceae
236.	*Peperomia bicolor*	-	Piperaceae
237.	*Peperomia caperata*	Emeral Ripple	Piperaceae
238.	*Peperomia clusiifolia*	Red-edged peperomia	Piperaceae
239.	*Peperomia cubensis*	-	Piperaceae
240.	*Peperomia glabella*	-	Piperaceae
241.	*Peperomia humilis*	-	Piperaceae
242.	*Peperomia magnoliaefolia*	-	Piperaceae
243.	*Peperomia maculosa*	Radiator plant	Piperaceae
244.	*Peperomia obtusifolia*	Baby rubber plant	Piperaceae
245.	*Peperomia orba*	-	Piperaceae
246.	*Peperomia rauhii*	-	Piperaceae
247.	*Peperomia sandersii*	Watermelon peperomia	Piperaceae
248.	*Peperomia scandens*	-	Piperaceae
249.	*Phalaris arundinacea*	Red Canary grass	Graminae

250.	*Philodendron bipinnatifidum*	-	Araceae
251.	*Philodendron erubescens*	-	Araceae
252.	*Philodendron fragrans*	-	Araceae
253.	*Philodendron imbe*	-	Araceae
254.	*Philodendron lacerum*	-	Araceae
255.	*Philodendron lacinatum*	-	Araceae
256.	*Philodendron mamei*	-	Araceae
257.	*Philodendron mandaianum*	-	Araceae
258.	*Philodendron panduraeforme*	-	Araceae
259.	*Philodendron pinnatifidum*	-	Araceae
260.	*Philodendron selloum*	-	Araceae
261.	*Philodendron sellowianum*	-	Araceae
262.	*Philodendron undulatum*	-	Araceae
263.	*Philodendron wend-imbe*	-	Araceae
264.	*Philodendron wendlandii*	-	Araceae
265.	*Philodendron williamsii*	-	Araceae
266.	*Pilea cadierei*	Aluminium plant	Urtiaceae
267.	*Pilea depressa*	-	Urtiaceae
268.	*Pilea macrophylla*	Artillary plant	Urtiaceae
269.	*Pilea serpyllacea*	-	Urtiaceae
270.	*Pleomele angustifolia honoriae*	-	Liliaceae

271.	*Plemele reflexa*	Song of India	Liliaceae
272.	*Rhoeo spathacea*	-	Commelinaceae
273.	*Ruellia devosiana*	-	Acanthaceae
274.	*Ruellia makoyana*	Monkey plant	Acanthaceae
275.	*Saintpaulia ionantha*	African violet	Gesneriaceae
276.	*Saintpaulia rupicola*	-	Gesneriaceae
277.	*Sansevieria cylindrica*	-	Liliaceae
278.	*Sansevieria desertii*	Rhino grass	Liliaceae
279.	*Sansevieria kirki pulchra*	-	Liliaceae
280.	*Sansevieria guineensis*	-	Liliaceae
281.	*Sansevieria trifasciata*	Snake plant	Liliaceae
282.	*Schefflera arboricola*	-	Araliaceae
283.	*Schefflera delavayi*	-	Araliaceae
284.	*Schefflera farinosa*	-	Araliaceae
285.	*Schefflera venulosa*	-	Araliaceae
286.	*Scindapsus aureus*	Pothas	Araceae
287.	*Scindapsus pictus*	Silver vine	Araceae
288.	*Scindapsus purpurea*	-	Araceae
289.	*Scindapsus speciosa*	-	Araceae
290.	*Sinningia speciosia fyfiana*	Florists's gloxinia	Gesneriaceae
291.	*Spathiphyllum cannaefolium*	-	Araceae

292.	*Spathiphyllum clevelandii*	-	Araceae
293.	*Spathiphyllum wallissi*	-	Araceae
294.	*Stanotaphrum secandatum*	St. Augustin grass	Graminae
295.	*Stromanthe sanguine*	-	Marantaceae
296.	*Syngonium macrophyllum*	-	Araceae
297.	*Syngonium podophyllum*	-	Araceae
298.	*Syngoinum wendlandi*	-	Araceae
299.	*Tacca chantrieri*	-	Taccaceae
300.	*Tacca cristata*	Bat flower	Taccaceae
301.	*Tillandsia cyanea*	-	Bromeliaceae
302.	*Tillandsia fendleri*	-	Bromeliaceae
303.	*Tillandsia flabellate*	-	Bromeliaceae
304.	*Tillandsia jaslisco-monticola*	-	Bromeliaceae
305.	*Tillandsia lindenii*	Blue flowered torch	Bromeliaceae
306.	*Tillandsia stricta*	-	Bromeliaceae
307.	*Tillandsia tricolor*	-	Bromeliaceae
308.	*Tradescantia albiflora 'Albovittata'*	-	Commelinaceae
309.	*Tradescantia blossfeldiana 'Variegata'*	-	Commelinaceae
310.	*Tradescantia fluminensis*	-	Commelinaceae
311.	*Tradescantia sillamontana*	-	Commelinaceae
312.	*Tupidanthus calyptratus*	-	Araliaceae

313.	*Vriesea bituminosa*	-	Bromeliaceae
314.	*Vriesea incurvata*	-	Bromeliaceae
315.	*Vriesea poelmanii*	-	Bromeliaceae
316.	*Vriesea splendens*	Flaming sword	Bromeliaceae
317.	*Xanthosoma jacquinii lineatum*	-	Araceae
318.	*Xanthosoma lindenii*	-	Araceae
319.	*Xanthosoma maculatum*	-	Araceae
320.	*Xanthosoma roseum*	-	Araceae
321.	*Zebrina pendula*	Silvery wandering jew	Commelinaceae

Chapter **18**

Computer Aided Design

I. General Information

Use of computers into the design process has benefited landscaping industry in several ways. There are several advantages in using computers for design. The most important factor is that the computer can store and handle large amount of engineering data, which can be used in various stages of design, development and manufacturer, like conceptual design, analysis, component design, documentation, process planning and manufacturing. The product development approach using the computer aided design and computer aided manufacturing (CAD/CAM) results in better and consistently good quality products invariably at a lower cost, with the attendant benefits of faster development of different variants of the product or model. The following table highlights the relative advantage of human designers and computers.

S.No.	Human designers are good at	Computers are good at
1.	Identifying design needs	Carry out long complex and laborious calculations
2.	Drawing up specifications	Store and efficiently search large data base
3.	Selecting design variables	Provide several alternative design solutions
4.	Optimizing design solutions	Provide information on design methodologies, heuristic data and stored expertise.
5.	Selecting the best design	Simulation of design for optimization
6.	Learning from experience	Aids in modeling, drafting and documentation

The wide acceptance and application of computers in design industry, is mainly due to the following developments during the past few years.

a. Development of interactive graphic and analysis software which prompts the user with the quarries and information on the video display unit and guides the user to solve the problems. This eliminates the need for processing in depth knowledge in high level programming languages.

b. Development of graphic workstations with powerful graphic facilities.

c. Availability of computers with enhanced memory and faster computing speed at affordable prices.

d. Reduction in product life cycle. e. Increasing design complexity.

Computer Aided Design

CAD technology makes use of the computer to created drawings of parts and assemblies on computer files which can be further analyzed and optimized. The functional, ergonomic and aesthetic features of the product can be evaluated on the computers. This has been made possible through the use of design workstations or CAD terminals and graphics and analytic software which help the designer to interactively model and analyze objects or components. The following lists the advantages of CAD.

1. Create conceptual model
2. Editing or refining the model to improve aesthetics, ergonomics and performance
3. Display the product in several colours to select colour combinations more appealing to customers.
4. Rotate and view the objects from various sides and directions.
5. Create and display all inner details of the assembly.
6. Check for interference and clearance between mating parts in static and dynamic situations.
7. Observe functional aspects of relative movements of various elements or assemblies.
8. Analyze stress, static deflection and dynamic behavior for different mechanical and thermal loading applications.

9. Study the object from various aspects such as material requirements, cost, value engineering manufacturing processes, standardization, simplification, variety reduction, service life, lubrication, servicing and maintenance aspects etc.
10. Prepare detailed component drawings giving full details of dimensions, tolerances, surface finish requirements, functional specifications etc.
11. Prepare assembly drawings depicting the orientation of components, assembly procedures and requirements and incorporation as required.
12. Prepare exploded view of assemblies for service and maintenance manuals. These views could be oriented to provide better visibility and improved comprehension of the design.
13. Plot/print the picture or drawing stores in computer file or the computer screen on different media.
14. Store the database of the object, part or drawing in a magnetic tape or CD for retrieval at a later date for use in some other design.

Point specification methods in Auto CAD

1. Absolute co-ordinate system

   The points are specified in x,y co-ordinate systems.

2. Relative co-ordinate system

   The points are specified with respect to the relation with the previous point with x & y distances i.e @ x displacement and y displacement.

3. Polar co-ordinate system

   The points are specified with respect to the relation with the previous point with displacement and angle i.e distance@<angle.

Limits of usage space in Auto CAD

A4 - 210x297 mm

A3 - 420x297 mm

A2 - 574x420 mm

A1 - 840x594 mm

A0 - 1188x594 mm

Units used in AutoCAD

1. Scientific – 1.55E+01
2. Decimal - 15.5
3. Engineering – 1-3.50′
4. Architectural – 1-31/2
5. Fractional - 151/2

DRAWING COMMANDS

Command: limits

Specify lower left corner or [ON/OFF] <0.0000,0.0000>:

Specify upper right corner <420.0000,297.0000>:

Command: z

ZOOM

Specify corner of window, enter a scale factor (nX or nXP), or [All/Center/Dynamic/Extents/Previous/Scale/Window] <real time>: a Regenerating model.

Command: l

LINE Specify first point: 50,50

Specify next point or [Undo]: 60,80

Specify next point or [Undo]: 90,100

Specify next point or [Close/Undo]:

Command: _mline

Current settings: Justification = Top, Scale = 20.00, Style = STANDARD Specify start point or [Justification/Scale/STyle]: 20,25

Specify next point: 80,55

Specify next point or [Undo]: 30,45

Specify next point or [Close/Undo]: 90,86

Specify next point or [Close/Undo]: 100,125

Specify next point or [Close/Undo]: 150,162

Specify next point or [Close/Undo]:

Command: _pline

Specify start point: 80,90

Current line-width is 0.0000

Specify next point or [Arc/Close/Halfwidth/Length/Undo/Width]: 78,90

Specify next point or [Arc/Close/Halfwidth/Length/Undo/Width]: a Specify endpoint of arc or [Angle/CEnter/CLose/Direction/Halfwidth/Line/Radius/Second pt/Undo/Width]: 100,150

Specify endpoint of arc or

[Angle/CEnter/CLose/Direction/Halfwidth/Line/Radius/Second pt/Undo/Width]: l

Specify next point or [Arc/Close/Halfwidth/Length/Undo/Width]: 156,135

Specify next point or [Arc/Close/Halfwidth/Length/Undo/Width]:

Command: _polygon

Enter number of sides <4>: 6

Specify center of polygon or [Edge]: 150,150

Enter an option [Inscribed in circle/Circumscribed about circle] <I>: Specify radius of circle: 98

Command: _rectangle

Specify first corner point or [Chamfer/Elevation/Fillet/Thickness/Width]: 189,200

Specify other corner point: 200,250

Command: _arc

Specify start point of arc or [CEnter]: 80,90

Specify second point of arc or [CEnter/ENd]: 90,60 Specify end point of arc: 50,45

Command: _circle

Specify center point for circle or [3P/2P/Ttr (tan tan radius)]: 250,250 Specify radius of circle or [Diameter]: 200

Command: _ellipse

Specify axis endpoint of ellipse or [Arc/Center]: 200,200

Specify other endpoint of axis: 300,200

Specify distance to other axis or [Rotation]: 30

EDITING COMMANDS

Command: _erase

Select objects: ALL 9 found

Command: _copy

Select objects: 1 1 found Specify base point or displacement, or [Multiple]: Specify second point of displacement or <use first point as displacement>:

Command: _mirror

Select objects: 1 found Specify first point of mirror line: Specify second point of mirror line: Delete source objects? [Yes/No] <N>:

Command: _offset

Specify offset distance or [Through] <1.0000>: 10

Select object to offset or <exit>: Specify point on side to offset: Select object to offset or <exit>: Specify point on side to offset: Select object to offset or <exit>:

Command: _array

Select objects: 1 found Enter the type of array [Rectangular/Polar] <R>:

Enter the number of rows (—) <1>: 4 Enter the number of columns (| | |) <1> 4 Enter the distance between rows or specify unit cell (—): 20 Specify the distance between columns (| | |): 20

Command: _array

Select objects: 1 found Enter the type of array [Rectangular/Polar] <R>: p Specify center point of array: Enter the number of items in the array: 8 Specify the angle to fill (+=ccw, -=cw) <360>: Rotate arrayed objects? [Yes/No] <Y>:

Command: _move

Select objects: 1 found Specify base point or displacement: Specify second point of displacement or <use first point as displacement>:

Command: m MOVE

Select objects: 1 found Specify base point or displacement: Specify second point of displacement or <use first point as displacement>:

Command: _rotate

Current positive angle in UCS: ANGDIR=counterclockwise ANGBASE=0 Select objects: 1 found Specify base point: Specify rotation angle or [Reference]: 56

Command: _scale

Select objects: 1 found Specify base point: Specify scale factor or [Reference]: 2

Command: _stretch

Select objects to stretch by crossing-window or crossing-polygon... Select objects: 1 found Specify base point or displacement: Specify second point of displacement:

Command: _lengthen

Select an object or [DElta/Percent/Total/DYnamic]: Current length: 494.4100 Select an object or [DElta/Percent/Total/DYnamic]:

Command: _trim

Current settings: Projection=UCS Edge=Extend Select cutting edges ... Select objects: 1 found Select object to trim or [Project/Edge/Undo]: Select object to trim or [Project/Edge/Undo]:

Command: _extend

Current settings: Projection=UCS Edge=Extend Select boundary edges ...

Select objects: 1 found Select object to extend or [Project/Edge/Undo]: Select object to extend or [Project/Edge/Undo]:

Command: _break

Select object: Specify second break point or [First point]:

Command: _chamfer (TRIM mode)

Current chamfer Dist1 = 10.0000, Dist2 = 10.0000 Select first line or [Polyline/Distance/Angle/Trim/Method]: Select second line:

Command: _fillet

Current settings: Mode = TRIM, Radius = 10.0000 Select first object or [Polyline/Radius/Trim]: Select second object:

II. Choose the best answer

1. The command that converts objects that enclose an area into 2D object ....

   Region Explode Revole

2. ....command refreshes the entire drawing from current view point

   Refresh Regen Pan

3. .... command creates linear objects that continues to infinity level

   Rectangular array Ray Stretch

4. .... command removes the un used items from the drawing

   Pyramid Delete Purge

5. ....is useful in producing a very large number of lines and points in the objects

   Hatching Poly line Array

6. ....command will be convenient to create a gap in objects

   Break Erase Trim

7. .... is a faster way for creating a line between two non parallel line

   Poly line Chamfer Line

8. .... is a closed geometric figure with equal sides and equal angles

   Circle   Arc   Regular polygone

9. Cannot terminate the line command by pressing which of the following key on the key board? ....

   Space bar   Back space   enter

10. To specify a point in a plane, take two mutually perpendicular lines as reference.....................

   Coordinate system   Dynamic input   Both

11. Which lines are used as reference to draw objects in AUTOCAD ....

   Ortho   Polar   Grid

12. .... keys helps to redo the previous actions

   CTRL + x   CTRL + y   CTRL + z

13. Which command is used to change the size of an existing object with respect to an existing entity?.....................

   Rotate   Move   Scale

14. .... is the arrangement of the objects around a point in a circular fashion

   Rectangular array   Polar array   a & b

15. The method of creating multiple copies of the selected objects and arranging them in a rectangular (or) circular fashion is called ....

   Copy   Hatch   Array

16. .... the sharp corners is another method of reducing the areas of stress concentration in the design

   Fillet   Elipse   Chamfering

17. Which of the following commands draws a line anywhere in 3D space, which starts from a point that specify and with the other end extending to infinity?

   PLINE   RAY   XLINE

18. ....option creator x lines that are parallel to the selected line x lines at a specified distance

   Fillet   Offset   Copy

19. Which of the following commands is used to exit from the AUTOCAD program?......................

Close End Quit

20. .... is a drawing file with pre established setting for new drawing

Display Template Window

21. ....object to meet the edges of other objects

Move Cut Trim

22. .... command is set to repeat the objects at different places of the workspace

Move Copy Mirror

III. Fill in the blanks

1. Usage space for A4 size in Auto CAD is ....mm
2. Fractional unit used in Auto CAD is ....
3. In absolute co-ordinate system, points are specified in ....systems
4. Types of point specification methods used in Auto CAD are............................
5. Command used in Auto CAD for zoom the picture is ....
6. Auto CAD creates a ....model
7. CAD is a ....
8. In polar co-ordinate system, points are specified with respect to the relation with the previous point with ....
9. In relative co-ordinate system, points are specified with respect to the relation with the previous point with ....
10. Usage space for A3 size in Auto CAD is ....mm
11. Usage space for A2 size in Auto CAD is ....mm
12. Usage space for A1 size in Auto CAD is ....mm
13. Usage space for A0 size in Auto CAD is ....mm
14. Engineering unit used in Auto CAD is ....
15. Scientific used in Auto CAD is ....

Answers

II. Choose the best answer

1. Region
2. Regen
3. Ray
4. Purge
5. Hatching
6. Break
7. Chamfer
8. Regular polygone
9. Space bar
10. Coordinate system
11. Grid
12. CTRL + y
13. Scale
14. Polar array
15. Array
16. Chamfering
17. PLINE
18. Offset
19. Close
20. Template
21. Trim
22. Copy

III. Fill in the blanks

1. 210 x 297
2. 151/2
3. x and y co-ordinate
4. Absolute co-ordinate system, relative co-ordinate system and polar co-ordinate system
5. z
6. conceptual
7. computer aided design
8. displacement and angle
9. x and y distances
10. 420 x 297
11. 574 x 420
12. 840 x 594
13. 1188 x 594
14. 1-3.50′
15. 1.55 E+01